Springer Tracts in Modern Physics
Volume 162

Managing Editor: G. Höhler, Karlsruhe

Editors: J. Kühn, Karlsruhe
Th. Müller, Karlsruhe
R. D. Peccei, Los Angeles
F. Steiner, Ulm
J. Trümper, Garching
P. Wölfle, Karlsruhe

Honorary Editor: E. A. Niekisch, Jülich

W0245849

Springer-Verlag Berlin Heidelberg GmbH

Springer Tracts in Modern Physics

Springer Tracts in Modern Physics provides comprehensive and critical reviews of topics of current interest in physics. The following fields are emphasized: elementary particle physics, solid-state physics, complex systems, and fundamental astrophysics.
Suitable reviews of other fields can also be accepted. The editors encourage prospective authors to correspond with them in advance of submitting an article. For reviews of topics belonging to the above mentioned fields, they should address the responsible editor, otherwise the managing editor.
See also http://www.springer.de/phys/books/stmp.html

Managing Editor

Gerhard Höhler

Institut für Theoretische Teilchenphysik
Universität Karlsruhe
Postfach 69 80
76128 Karlsruhe, Germany
Phone: +49 (7 21) 6 08 33 75
Fax: +49 (7 21) 37 07 26
Email: gerhard.hoehler@physik.uni-karlsruhe.de
http://www-ttp.physik.uni-karlsruhe.de/

Elementary Particle Physics, Editors

Johann H. Kühn

Institut für Theoretische Teilchenphysik
Universität Karlsruhe
Postfach 69 80
76128 Karlsruhe, Germany
Phone: +49 (7 21) 6 08 33 72
Fax: +49 (7 21) 37 07 26
Email: johann.kuehn@physik.uni-karlsruhe.de
http://www-ttp.physik.uni-karlsruhe.de/~jk

Thomas Müller

Institut für Experimentelle Kernphysik
Fakultät für Physik
Universität Karlsruhe
Postfach 69 80
76128 Karlsruhe, Germany
Phone: +49 (7 21) 6 08 35 24
Fax: +49 (7 21) 6 07 26 21
Email: thomas.muller@physik.uni-karlsruhe.de
http://www-ekp.physik.uni-karlsruhe.de

Roberto Peccei

Department of Physics
University of California, Los Angeles
405 Hilgard Avenue
Los Angeles, CA 90024-1547, USA
Phone: +1 310 825 1042
Fax: +1 310 825 9368
Email: peccei@physics.ucla.edu
http://www.physics.ucla.edu/faculty/ladder/
peccei.html

Solid-State Physics, Editor

Peter Wölfle

Institut für Theorie der Kondensierten Materie
Universität Karlsruhe
Postfach 69 80
76128 Karlsruhe, Germany
Phone: +49 (7 21) 6 08 35 90
Fax: +49 (7 21) 69 81 50
Email: woelfle@tkm.physik.uni-karlsruhe.de
http://www-tkm.physik.uni-karlsruhe.de

Complex Systems, Editor

Frank Steiner

Abteilung Theoretische Physik
Universität Ulm
Albert-Einstein-Allee 11
89069 Ulm, Germany
Phone: +49 (7 31) 5 02 29 10
Fax: +49 (7 31) 5 02 29 24
Email: steiner@physik.uni-ulm.de
http://www.physik.uni-ulm.de/theo/theophys.html

Fundamental Astrophysics, Editor

Joachim Trümper

Max-Planck-Institut für Extraterrestrische Physik
Postfach 16 03
85740 Garching, Germany
Phone: +49 (89) 32 99 35 59
Fax: +49 (89) 32 99 35 69
Email: jtrumper@mpe-garching.mpg.de
http://www.mpe-garching.mpg.de/index.html

Wolfgang Hollik   Günter Duckeck

Electroweak Precision Tests at LEP

With 60 Figures

 Springer

Professor Dr. Wolfgang Hollik
Universität Karlsruhe
Institut für Theoretische Physik
Kaiserstrasse 12
76131 Karlsruhe, Germany
E-mail: hollik@particle.physik.uni-karlsruhe.de

Dr. Günter Duckeck
Ludwig-Maximilians-Universität München
Sektion Physik
Am Coulombwall 1
85748 Garching, Germany
E-mail: guenter.duckeck@physik.uni-muenchen.de

Library of Congress Cataloging-in-Publication Data applied for.

Die Deutsche Bibliothek – CIP Einheitsaufnahme

Hollik, Wolfgang: Electroweak precision tests at LEP / Wolfgang Hollik; Günter Duckeck. – Berlin; Heidelberg; New York; Barcelona; Hong Kong; London; Milan; Paris; Singapore; Tokyo: Springer, 2000 (Springer tracts in modern physics; Vol. 162) ISBN 3-540-66541-2

Physics and Astronomy Classification Scheme (PACS): 12.15.Ji, 12.15.Lk, 13.38.Dg, 14.70.-e, 14.80.Bn

ISSN 0081-3869
ISBN 978-3-662-15655-1   ISBN 978-3-540-48163-8 (eBook)
DOI 10.1007/978-3-540-48163-8

© Springer-Verlag Berlin Heidelberg 2000
Originally published by Springer-Verlag Berlin Heidelberg New York in 2000.
Softcover reprint of the hardcover 1st edition 2000

Typesetting: Camera-ready copy by the author using a Springer LaTeX macro package
Cover design: *design & production* GmbH, Heidelberg

Printed on acid-free paper   SPIN: 10572651   56/3144/tr   5 4 3 2 1 0

Preface

The present generation of high-precision experiments imposes stringent tests on the Standard Model. Besides the impressive achievements in the determination of the Z^0 boson parameters and the W mass, the most important step has been the discovery of the top quark at the TEVATRON with a mass determination, now at $m_t = 174.3 \pm 5.1$ GeV, which coincides with the mass range indirectly obtained via the radiative corrections. Moreover, with the top mass as an additional precise experimental data point one can now fully exploit the sensitivity to the Higgs mass.

The high experimental sensitivity in the electroweak observables, at the level of the quantum effects, requires the highest standards for both theory and experiment. The current particle physics experiments explore electron–positron and proton–antiproton collisions in a new range of centre-of-mass energies. The large data volumes recorded in recent years and the employment of new detector technologies provide for measurements with unprecedented precision. On the theoretical side a sizeable amount of work has contributed over the last few years to a steadily rising improvement of the Standard Model predictions, pinning down the theoretical uncertainties to a level sufficiently small for the current interpretation of the precision data, but still sizeable enough to cause problems in a further increase in the experimental accuracy.

The lack of direct signals from "new physics" makes the high-precision experiments also a unique tool in the search for *indirect* effects, through definite deviations of the experimental results from the theoretical predictions of the minimal Standard Model. Since such deviations are expected to be small, of the typical size of the Standard Model radiative corrections, it is essential to have the standard loop effects in the precision observables under control.

In the first chapter of this book we give a brief discussion of the structure of the Standard Model and its quantum corrections for testing the electroweak theory at present and future colliders. The predictions for the vector boson masses, the neutrino scattering cross-sections and the Z^0 resonance observables, such as the width of the Z^0 resonance, partial widths, effective neutral-current coupling constants and mixing angles at the Z^0 peak, are presented in some detail. Finally we address the question of how virtual new physics can influence the predictions of the precision observables and discuss the minimal supersymmetric Standard Model (MSSM) as a special example of particular theoretical interest.

In the second chapter the experimental measurements of the electroweak observables are described. The main focus is put on the measurements performed at the e^+e^- collider LEP at energies close to the Z^0 peak. From the cross-sections and asymmetries for the various final states the properties of the Z^0, its mass, decay width and couplings can be precisely determined. In addition we present the complementary measurements at the SLC e^+e^- collider and summarize the status of the ongoing LEP measurements at energies above the W^+W^- pair production threshold. Finally we briefly review electroweak results from other experiments, the mass measurements of the top quark and the W boson at the TEVATRON and the determination of the electroweak mixing angle in neutrino–nucleon scattering.

In the third chapter we compare the theoretical predictions with the experimental data and discuss their implications for the present status of the Standard Model.

Successful collaborations and discussions helpful to shaping the theoretical description of the electroweak precision observables in the Standard Model and the MSSM are gratefully acknowledged, with special thanks to Dima Bardin, Wim de Boer, Andreas Dabelstein, Giuseppe Degrassi, Jens Erler, Sven Heinemeyer, Fred Jegerlehner, Hans Kühn, Giampiero Passarino, Ulrich Schwickerath and Georg Weiglein. The experimental measurements reviewed in this book are the result of the work of many hundreds of physicists at LEP and SLC over the past years. Special thanks go to the members of the LEP electroweak working group for their effort in combining the precision electroweak measurements and, in particular, to Dorothee Schaile, Günter Quast and Tatsuo Kawamoto for useful discussions on the material presented here.

Karlsruhe – München, October 1999   *Wolfgang Hollik*
                                                     Günter Duckeck

Contents

1. Theory of Precision Tests

The theory of the electroweak interaction, also known as the electroweak Standard Model [1–4], is a gauge-invariant quantum field theory with the symmetry group $\mathrm{SU}(2) \times \mathrm{U}(1)$ spontaneously broken by the Higgs mechanism. It contains essentially three free parameters to describe the gauge bosons $\gamma, W^\pm, Z$ and their interactions with the fermions. For a comparison between theory and experiment three independent experimental input data values are required. The most natural choice is given by the electromagnetic fine structure constant α, the muon decay constant (Fermi constant) G_μ and the mass of the Z^0 boson, which has now been measured with the same accuracy as G_μ. Other measurable quantities are then predicted in terms of the input data. Each additional precision experiment which allows the detection of small deviations from the lowest-order predictions can be considered a test of the electroweak theory at the quantum level. In the Feynman graph expansion of the scattering amplitude for a given process the higher-order terms show up as diagrams containing closed loops. The lowest-order amplitudes could also be derived from a corresponding classical field theory, whereas the loop contributions can only be obtained from the quantized version. The renormalizability of the Standard Model [5] ensures that it retains its predictive power in higher orders also. The higher-order terms, commonly called radiative corrections, are the quantum effects of the electroweak theory. They are complicated in their concrete form, but they are ultimately the consequence of a basic Lagrangian with a simple structure. The quantum corrections contain the self-coupling of the vector bosons as well as their interactions with the Higgs field and the top quark, and provide the theoretical basis for electroweak precision tests. Assuming the validity of the Standard Model, the presence of the top quark and the Higgs boson in the loop contributions to electroweak observables allows one to obtain significant bounds on their masses from precision measurements of these observables. An important step strengthening our confidence in the Standard Model has been the discovery of the top quark at the TEVATRON [6] at a mass that agrees with the mass range obtained indirectly from the radiative corrections.

1.1 The Electroweak Standard Model

The empirical knowledge about the structure of the electroweak interaction of
the fundamental constituents of matter can be embedded in a gauge-invariant
field theory of the unified electromagnetic and weak interactions by interpreting SU(2)×U(1) as the group of local gauge transformations under which the
Lagrangian is invariant. This full symmetry has to be broken by the Higgs
mechanism down to the electromagnetic gauge symmetry; otherwise the $W^{\pm}$
and Z^0 bosons would be massless. The minimal formulation, the Standard
Model, requires a single scalar field, the Higgs field, which is a doublet under
SU(2).

1.1.1 The Standard Model Lagrangian

The phenomenological basis for the formulation of the Standard Model is
given by the following empirical facts.

- The SU(2)×U(1) family structure of the fermions. The fermions appear as
  families with left-handed doublets and right-handed singlets:

$$\begin{pmatrix} \nu_e \\ e \end{pmatrix}_{\mathrm{L}}, \quad \begin{pmatrix} \nu_\mu \\ \mu \end{pmatrix}_{\mathrm{L}}, \quad \begin{pmatrix} \nu_\tau \\ \tau \end{pmatrix}_{\mathrm{L}}, \quad e_{\mathrm{R}}, \quad \mu_{\mathrm{R}}, \quad \tau_{\mathrm{R}}$$

$$\begin{pmatrix} u \\ d \end{pmatrix}_{\mathrm{L}}, \quad \begin{pmatrix} c \\ s \end{pmatrix}_{\mathrm{L}}, \quad \begin{pmatrix} t \\ b \end{pmatrix}_{\mathrm{L}}, \quad u_{\mathrm{R}}, \quad d_{\mathrm{R}}, \quad c_{\mathrm{R}}, \cdots$$

  They can be characterized by the quantum numbers of the weak isospin I
  and I_3 and the weak hypercharge Y.
- The Gell-Mann–Nishijima relation. Between the quantum numbers classi-
  fying the fermions with respect to the group SU(2)×U(1) and their electric
  charges Q, the relation

$$Q = I_3 + \frac{Y}{2} \tag{1.1}$$

  is valid.
- The existence of vector bosons. There are four vector bosons as carriers of
  the electroweak force:

$$\gamma, \quad W^+, \quad W^-, \quad Z^0,$$

  where the photon is massless and the $W^{\pm}$ and Z^0 have masses m_W, m_Z
  different from 0.

According to the general principles of constructing a gauge-invariant field
theory with spontaneous symmetry breaking, the gauge, Higgs and fermion
parts of the electroweak Lagrangian at the classical level

$$\mathcal{L}_{\mathrm{cl}} = \mathcal{L}_{\mathrm{G}} + \mathcal{L}_{\mathrm{H}} + \mathcal{L}_{\mathrm{F}} \tag{1.2}$$

are specified in the following way.

Gauge Fields. $SU(2) \times U(1)$ is a non-Abelian group which is generated by the isospin operators I_1, I_2, I_3 and the hypercharge Y (the elements of the corresponding Lie algebra). Each of these generalized charges is associated with a vector field: a triplet of vector fields $W_\mu^{1,2,3}$ with $I_{1,2,3}$ and a singlet field B_μ with Y. The isotriplet W_μ^a, $a = 1, 2, 3$, and the isosinglet B_μ lead to the field strength tensors

$$
\begin{aligned}
W_{\mu\nu}^a &= \partial_\mu W_\nu^a - \partial_\nu W_\mu^a + g_2 \, \epsilon_{abc} \, W_\mu^b W_\nu^c, \\
B_{\mu\nu} &= \partial_\mu B_\nu - \partial_\nu B_\mu.
\end{aligned}
\tag{1.3}
$$

g_2 denotes the non-Abelian $SU(2)$ gauge coupling constant and g_1 the Abelian $U(1)$ coupling. From the field tensors (1.3) the pure gauge field Lagrangian

$$
\mathcal{L}_G = -\frac{1}{4} \, W_{\mu\nu}^a W^{\mu\nu,a} - \frac{1}{4} \, B_{\mu\nu} B^{\mu\nu}
\tag{1.4}
$$

is formed according to the rules for the non-Abelian case.

Fermion Fields and Fermion–Gauge Interaction. The left-handed fermion fields of each lepton and quark family (the colour index is suppressed),

$$
\psi_j^{\mathrm{L}} = \begin{pmatrix} \psi_{j+}^{\mathrm{L}} \\ \psi_{j-}^{\mathrm{L}} \end{pmatrix}, \quad \psi^{\mathrm{L}} \equiv \frac{1 - \gamma_5}{2} \psi,
$$

with family index j are grouped into $SU(2)$ doublets with component index $\sigma = \pm$, and the right-handed fields into singlets

$$
\psi_j^{\mathrm{R}} = \psi_{j\sigma}^{\mathrm{R}}, \quad \psi^{\mathrm{R}} \equiv \frac{1 + \gamma_5}{2} \psi.
$$

Each left- and right-handed multiplet is an eigenstate of the weak hypercharge Y such that the relation (1.1) is fulfilled. The covariant derivative

$$
D_\mu = \partial_\mu - i g_2 \, I_a W_\mu^a + i g_1 \frac{Y}{2} \, B_\mu
\tag{1.5}
$$

induces the fermion–gauge field interaction via the minimal substitution rule:

$$
\mathcal{L}_F = \sum_j \bar{\psi}_j^{\mathrm{L}} i \gamma^\mu D_\mu \psi_j^{\mathrm{L}} + \sum_{j,\sigma} \bar{\psi}_{j\sigma}^{\mathrm{R}} i \gamma^\mu D_\mu \psi_{j\sigma}^{\mathrm{R}}.
\tag{1.6}
$$

Higgs Field, Higgs–Gauge and Yukawa Interactions. For spontaneous breaking of the $SU(2) \times U(1)$ symmetry leaving the electromagnetic gauge subgroup $U(1)_{\mathrm{em}}$ unbroken, a single complex scalar doublet field with hypercharge $Y = 1$

$$
\Phi(x) = \begin{pmatrix} \phi^+(x) \\ \phi^0(x) \end{pmatrix}
\tag{1.7}
$$

is coupled to the gauge fields according to the Lagrangian

$$\mathcal{L}_{\mathrm{H}} = (D_\mu \Phi)^+ (D^\mu \Phi) - V(\Phi) \qquad (1.8)$$

with the covariant derivative

$$D_\mu = \partial_\mu - i\, g_2\, I_a W_\mu^a + i\frac{g_1}{2} B_\mu . \qquad (1.9)$$

The Higgs field self-interaction

$$V(\Phi) = -\mu^2\, \Phi^+ \Phi + \frac{\lambda}{4}\, (\Phi^+ \Phi)^2 \qquad (1.10)$$

is constructed in such a way that it has a non-vanishing vacuum expectation value v, related to the coefficients of the potential V by

$$v = \frac{2\mu}{\sqrt{\lambda}} . \qquad (1.11)$$

The field (1.7) can be written in the following way:

$$\Phi(x) = \begin{pmatrix} \phi^+(x) \\ [v + H(x) + i\chi(x)]/\sqrt{2} \end{pmatrix}, \qquad (1.12)$$

where the components ϕ^+, H, χ now have vacuum expectation values of zero. Exploiting the invariance of the Lagrangian, one notices that the components ϕ^+, χ can be gauged away, which means that they are unphysical (Higgs ghosts or would-be Goldstone bosons). In this particular gauge, the unitary gauge, the Higgs field has the simple form

$$\Phi(x) = \frac{1}{\sqrt{2}} \begin{pmatrix} 0 \\ v + H(x) \end{pmatrix} . \qquad (1.13)$$

The real part of ϕ^0, $H(x)$, is the field of a physical neutral scalar particle, the Higgs boson, with mass

$$m_H = \mu\sqrt{2}. \qquad (1.14)$$

The Higgs field components have triple and quartic self-couplings following from V, and couplings to the gauge fields via the kinetic term of (1.8).

In addition, Yukawa couplings to fermions are introduced in order to make the charged fermions massive. The Yukawa term is conveniently expressed in the doublet field components (1.7). For one family of leptons and quarks it reads

$$\begin{aligned}
\mathcal{L}_{\mathrm{Yukawa}} = &-g_l \left(\bar{\nu}_{\mathrm{L}}\, \phi^+\, l_{\mathrm{R}} + \bar{l}_{\mathrm{R}}\, \phi^-\, \nu_{\mathrm{L}} + \bar{l}_{\mathrm{L}}\, \phi^0\, l_{\mathrm{R}} + \bar{l}_{\mathrm{R}}\, \phi^{0*}\, l_{\mathrm{L}} \right) \\
&-g_d \left(\bar{u}_{\mathrm{L}}\, \phi^+\, d_{\mathrm{R}} + \bar{d}_{\mathrm{R}}\, \phi^-\, u_{\mathrm{L}} + \bar{d}_{\mathrm{L}}\, \phi^0\, d_{\mathrm{R}} + \bar{d}_{\mathrm{R}}\, \phi^{0*}\, d_{\mathrm{L}} \right) \\
&+g_u \left(\bar{u}_{\mathrm{R}}\, \phi^+\, d_{\mathrm{L}} + \bar{d}_{\mathrm{L}}\, \phi^-\, u_{\mathrm{R}} - \bar{u}_{\mathrm{R}}\, \phi^0\, u_{\mathrm{L}} - \bar{u}_{\mathrm{L}}\, \phi^{0*}\, u_{\mathrm{R}} \right). \quad (1.15)
\end{aligned}$$

ϕ^- denotes the adjoint of ϕ^+.

Because $v \neq 0$ fermion mass terms are induced. The Yukawa coupling constants $g_{l,d,u}$ are related to the masses of the charged fermions by (1.31). In the unitary gauge the Yukawa Lagrangian is particularly simple:

$$\mathcal{L}_{\text{Yukawa}} = -\sum_f m_f \, \bar{\psi}_f \psi_f - \sum_f \frac{m_f}{v} \, \bar{\psi}_f \psi_f \, H \, . \tag{1.16}$$

As a remnant of this mechanism for generating fermion masses in a gauge invariant way, Yukawa interactions between the massive fermions and the physical Higgs field occur with coupling constants proportional to the fermion masses.

Physical Fields and Parameters. The gauge-invariant Higgs–gauge field interaction in the kinetic part of (1.8) gives rise to mass terms for the vector bosons in the non-diagonal form

$$\frac{1}{2} \left(\frac{g_2}{2} v \right)^2 (W_1^2 + W_2^2) + \frac{1}{2} \left(W_\mu^3, B_\mu \right) \frac{v^2}{4} \begin{pmatrix} g_2^2 & g_1 g_2 \\ g_1 g_2 & g_1^2 \end{pmatrix} \begin{pmatrix} W_\mu^3 \\ B_\mu \end{pmatrix} . \tag{1.17}$$

The physical content becomes transparent when one performs a transformation from the fields W_μ^a, B_μ (in terms of which the symmetry is manifest) to the physical fields

$$W_\mu^\pm = \frac{1}{\sqrt{2}} \left(W_\mu^1 \mp i W_\mu^2 \right) \tag{1.18}$$

and

$$\begin{aligned} Z_\mu &= \quad \cos\theta_W \, W_\mu^3 + \sin\theta_W \, B_\mu \, , \\ A_\mu &= -\sin\theta_W \, W_\mu^3 + \cos\theta_W \, B_\mu \, . \end{aligned} \tag{1.19}$$

In these fields the mass term (1.17) is diagonal and has the form

$$m_W^2 \, W_\mu^+ W^{-\mu} + \frac{1}{2} \left(A_\mu, Z_\mu \right) \begin{pmatrix} 0 & 0 \\ 0 & m_Z^2 \end{pmatrix} \begin{pmatrix} A^\mu \\ Z^\mu \end{pmatrix} \tag{1.20}$$

with

$$\begin{aligned} m_W &= \frac{1}{2} \, g_2 v \, , \\ m_Z &= \frac{1}{2} \sqrt{g_1^2 + g_2^2} \; v \, . \end{aligned} \tag{1.21}$$

The mixing angle in the rotation (1.19) is given by

$$\cos\theta_W = \frac{g_2}{\sqrt{g_1^2 + g_2^2}} = \frac{m_W}{m_Z} \, . \tag{1.22}$$

The fermion–gauge interaction term of the Standard Model Lagrangian contains the electromagnetic current (the summation extends over the fermion fields ψ_f)

$$J_\mu^{\text{em}} = \sum_f Q_f \, \bar{\psi}_f \gamma_\mu \psi_f \, , \tag{1.23}$$

the weak neutral current

$$J_\mu^Z = \sum_f \bar{\psi}_f \gamma_\mu (v_f - a_f \gamma_5) \psi_f \equiv 2 \left(J_\mu^L - \sin^2 \theta_W J_\mu^{em} \right) \qquad (1.24)$$

with the coupling constants

$$v_f = I_3^f - 2 Q_f \sin^2 \theta_W \,,$$
$$a_f = I_3^f \,, \qquad\qquad\qquad\qquad\qquad (1.25)$$

where Q_f and I_3^f denote the charge and the third isospin component of f_L, and the weak charged current

$$J_\mu^+ = \sum_{\ell=e,\mu,\tau} \bar{\nu}_\ell \gamma_\mu (1 - \gamma_5)\ell + (\bar{u}, \bar{c}, \bar{t})\gamma_\mu (1 - \gamma_5) U_{CKM} \begin{pmatrix} u \\ s \\ b \end{pmatrix} \qquad (1.26)$$

in the following way (h.c. denotes the hermitian conjugate):

$$\mathcal{L}_{f\text{-}G} = -e\, J_\mu^{em} A^\mu + \frac{g_2}{2\cos\theta_W} J_\mu^Z Z^\mu + \frac{g_2}{2\sqrt{2}} (J_\mu^+ W^{\mu-} + \text{h.c.})\,, \qquad (1.27)$$

with the Cabibbo–Kobayashi–Maskawa matrix U_{CKM} [3]

$$U_{CKM} = \begin{pmatrix} V_{ud} & V_{us} & V_{ub} \\ V_{cd} & V_{cs} & V_{cb} \\ V_{td} & V_{ts} & V_{tb} \end{pmatrix}\,, \qquad (1.28)$$

which describes family mixing in the quark sector. Its origin is the diagonalization of the quark mass matrices from the Yukawa coupling which appears since quarks of the same charge have different masses. For massless neutrinos no mixing in the leptonic sector is present. Owing to the unitarity of U_{CKM} the mixing is absent in the neutral current.

The electric fundamental charge e can be expressed in terms of the gauge couplings in the following way:

$$e = \frac{g_1 g_2}{\sqrt{g_1^2 + g_2^2}} \qquad (1.29)$$

or

$$g_2 = \frac{e}{\sin\theta_W}\,, \qquad g_1 = \frac{e}{\cos\theta_W}\,. \qquad (1.30)$$

Finally, from the Yukawa coupling terms in (1.15), the fermion masses are obtained:

$$m_f = g_f \frac{v}{\sqrt{2}} = \sqrt{2}\,\frac{g_f}{g_2}\, m_W\,. \qquad (1.31)$$

The relations above allow one to replace the original set of parameters

$$g_2,\ g_1,\ \lambda,\ \mu^2,\ g_f\,, \qquad (1.32)$$

by the equivalent set of more physical parameters

$$e,\ m_W,\ m_Z,\ m_H,\ m_f\,, \qquad (1.33)$$

where each of them can (in principle) directly be measured in a suitable experiment.

An additional very precisely measured parameter is the Fermi constant G_μ, which is the effective four-fermion coupling constant in the Fermi model, measured by the muon lifetime (see Sect. 1.3.1). Consistency of the Standard Model at $q^2 \ll m_W^2$ with the Fermi model requires the identification

$$\frac{G_\mu}{\sqrt{2}} = \frac{e^2}{8 \sin^2 \theta_W m_W^2} , \tag{1.34}$$

which allows us to relate the vector boson masses to the parameters α, G_μ and $\sin^2 \theta_W$ as follows:

$$m_W^2 = \frac{\pi \alpha}{\sqrt{2} G_\mu} \frac{1}{\sin^2 \theta_W} ,$$
$$m_Z^2 = \frac{\pi \alpha}{\sqrt{2} G_\mu} \frac{1}{\sin^2 \theta_W \cos^2 \theta_W} ; \tag{1.35}$$

and thus to establish also the m_W–m_Z interdependence:

$$m_W^2 \left(1 - \frac{m_W^2}{m_Z^2}\right) = \frac{\pi \alpha}{\sqrt{2} G_\mu} . \tag{1.36}$$

1.1.2 Gauge Fixing and Ghost Fields

Since the S-matrix element for any physical process is a gauge-invariant quantity it is possible to work in the unitary gauge with no unphysical particles in the internal stages. For a systematic treatment of the quantization of $\mathcal{L}_{cl}$ and for higher-order calculations, however, it is better to refer to a renormalizable gauge. This can be done by adding to $\mathcal{L}_{cl}$ a gauge-fixing Lagrangian, for example

$$\mathcal{L}_{\text{fix}} = -\frac{1}{2} \left(F_\gamma^2 + F_Z^2 + 2F_+ F_-\right) \tag{1.37}$$

with linear gauge fixings of the 't Hooft type,

$$F_\pm = \frac{1}{\sqrt{\xi^W}} \left(\partial^\mu W_\mu^\pm \mp i m_W \xi^W \phi^\pm\right) ,$$
$$F_Z = \frac{1}{\sqrt{\xi^Z}} \left(\partial^\mu Z_\mu - m_Z \xi^Z \chi\right) ,$$
$$F_\gamma = \frac{1}{\sqrt{\xi^\gamma}} \partial^\mu A_\mu , \tag{1.38}$$

with arbitrary parameters $\xi^{W,Z,\gamma}$. In this class of 't Hooft gauges, the vector boson propagators have the form

$$\frac{i}{q^2 - m_V^2} \left(-g^{\mu\nu} + \frac{(1 - \xi^V)q^\mu q^\nu}{q^2 - \xi^V m_V^2} \right)$$

$$= \frac{i}{q^2 - m_V^2} \left(-g^{\mu\nu} + \frac{q^\mu q^\nu}{q^2} \right) + \frac{i\xi^V}{q^2 - \xi^V m_V^2} \frac{q^\mu q^\nu}{q^2}, \qquad (1.39)$$

the propagators for the unphysical Higgs fields are given by

$$\frac{i}{q^2 - \xi^W m_W^2} \quad \text{for } \phi^\pm,$$

$$\frac{i}{q^2 - \xi^Z m_Z^2} \quad \text{for } \chi^0, \qquad\qquad (1.40)$$

and Higgs–vector boson transitions do not occur.

For completion of the renormalizable Lagrangian, the Faddeev–Popov ghost term $\mathcal{L}_{\text{gh}}$ has to be added [7] in order to balance the undesired effects of the unphysical components introduced by $\mathcal{L}_{\text{fix}}$:

$$\mathcal{L} = \mathcal{L}_{\text{cl}} + \mathcal{L}_{\text{fix}} + \mathcal{L}_{\text{gh}}, \qquad\qquad (1.41)$$

where

$$\mathcal{L}_{\text{gh}} = \bar{u}^\alpha(x) \frac{\delta F^\alpha}{\delta \theta^\beta(x)} u^\beta(x) \qquad\qquad (1.42)$$

with ghost fields u^γ, u^Z, $u^\pm$, and $\delta F^\alpha / \delta \theta^\beta$ being the change of the gauge-fixing operators (1.38) under infinitesimal gauge transformations characterized by the gauge functions $\theta^\alpha(x) = \{\theta^a(x), \theta^Y(x)\}$.

In the 't Hooft–Feynman gauge ($\xi = 1$) the vector boson propagators (1.39) become particularly simple: the transverse and longitudinal components, as well as the propagators for the unphysical Higgs fields $\phi^\pm$ and χ and the ghost fields $u^\pm$ and u^Z, have poles which coincide with the masses of the corresponding physical particles $W^\pm$ and Z^0.

1.1.3 Feynman Rules

We can write down the Lagrangian expressed in terms of the physical parameters,

$$\mathcal{L}(A_\mu, W_\mu^\pm, Z_\mu, H, \phi^\pm, \chi, u^\pm, u^Z, u^\gamma; \, m_W, m_Z, e, \ldots)$$

in a way which allows us to read off the propagators and the vertices directly. We specify the propagators in the $R_{\xi=1}$ gauge, where the vector boson propagators have the simple algebraic form $\sim g_{\mu\nu}$:

$$\mathcal{L}_{\text{G}} + \mathcal{L}_{\text{H}}$$

$$= \frac{1}{2} A_\mu \,\square\, A^\mu \quad \leftrightarrow \quad \frac{-i\, g_{\mu\nu}}{q^2}$$

$$+ W_\mu^- \left(\Box + m_W^2\right) W^{+\mu} \quad \leftrightarrow \quad \frac{-\mathrm{i}\, g_{\mu\nu}}{q^2 - m_W^2}$$

$$+ \frac{1}{2} Z_\mu \left(\Box + m_Z^2\right) Z^\mu \quad \leftrightarrow \quad \frac{-\mathrm{i}\, g_{\mu\nu}}{q^2 - m_Z^2}$$

$$+ \frac{1}{2} H \left(\Box + m_H^2\right) H \quad \leftrightarrow \quad \frac{\mathrm{i}}{q^2 - m_H^2}$$

$$+ \text{interaction terms} \qquad [VVV],\ [VVVV],$$
$$[VVH],\ [VVHH],$$
$$[HHH],\ [HHHH]$$

$$+ \text{ghost terms};$$

$$\mathcal{L}_\mathrm{F} + \mathcal{L}_\text{Yukawa}$$

$$= \sum_f \bar{f} \left(\mathrm{i}\, \slashed{\partial} - m_f\right) f \quad \leftrightarrow \quad \frac{\mathrm{i}}{\slashed{q} - m_f}$$

$$+ J_\text{em}^\mu\, A_\mu \quad \leftrightarrow \quad -\mathrm{i}\, e\, Q_f \gamma_\mu$$

$$+ J_\text{NC}^\mu\, Z_\mu \quad \leftrightarrow \quad \mathrm{i}\, \frac{e}{2 \sin\theta_W \cos\theta_W}\, \gamma_\mu (v_f - a_f \gamma_5)$$

$$+ J_\text{CC}^\mu\, W_\mu \quad \leftrightarrow \quad \mathrm{i}\, \frac{e}{2\sqrt{2} \sin\theta_W}\, \gamma_\mu (1 - \gamma_5)\, V_{jk}$$

$$- \frac{g_f}{\sqrt{2}}\, \bar{f} f H \quad \leftrightarrow \quad -\mathrm{i}\, \frac{g_f}{\sqrt{2}} = \mathrm{i}\, \frac{e}{2 \sin\theta_W}\, \frac{m_f}{m_W}$$

$$+ \cdots \tag{1.43}$$

Not listed are the fermion couplings to the unphysical Higgs ghosts χ, $\phi^\pm$. The quantities V_{jk} in the charged-current vertex are the elements of the CKM matrix (1.28). For the complete list of all interaction vertices we refer to the literature [8]. For the complete list of all interaction vertices we refer to the literature [8].

In order to describe scattering processes between light fermions in lowest order we can, in most cases, neglect the exchange of Higgs bosons because of their small Yukawa couplings to the light fermions. The standard processes accessible by the experimental facilities are basically four-fermion processes. These are mediated by the gauge bosons and are defined, sufficiently well in lowest order, by the vertices for the fermions interacting with the vector bosons. The Feynman rules given above thus provide the ingredients to calculate the lowest-order amplitudes for fermionic processes.

1.2 Renormalization of the Electroweak Parameters

The possibility of performing precision tests is based on the formulation of
the Standard Model as a renormalizable quantum field theory preserving
its predictive power beyond tree-level calculations. When the experimental
accuracy is sensitive to the loop-induced quantum effects, the Higgs sector
of the Standard Model is also being probed. The higher-order terms induce
a sensitivity of electroweak observables to the top and Higgs masses m_t, m_H
and to the strong coupling constant α_s.

1.2.1 Basic Strategy

In higher-order perturbation theory the relations between the formal parameters and measurable quantities are different from the tree-level relations in
general. Moreover, the procedure is obscured by the appearance of divergences in the loop integrations. For a mathematically consistent treatment
one has to regularize the theory, e.g. by dimensional regularization (performing the calculations in D dimensions). But then the relations between the
physical quantities and the parameters become cut-off-dependent. Hence, the
parameters of the basic Lagrangian, the "bare" parameters, have no physical meaning. On the other hand, the relations between measurable physical
quantities, where the parameters drop out, are finite and independent of the
cut-off. It is therefore in principle possible to perform tests of the theory in
terms of such relations by eliminating the bare parameters [9,10].

Alternatively, one may replace the bare parameters by renormalized ones
by multiplicative renormalization for each bare parameter a_0,

$$a_0 = Z_a\, a = a + \delta a \tag{1.44}$$

with renormalization constants Z_a different from 1 by a higher-order term.
The renormalized parameters a are finite and fixed by a set of renormalization
conditions. The decomposition (1.44) is to a large extent arbitrary. Only the
divergent parts are determined directly by the structure of the divergences of
the one-loop amplitudes. The finite parts depend on the choice of the explicit
renormalization conditions.

The simplest way to obtain a set of finite Green functions is the "minimal
subtraction scheme" [11], where (in dimensional regularization) the singular
part of each divergent diagram is subtracted and the parameters are defined
at an arbitrary mass scale μ. This scheme, with slight modifications, has been
applied in QCD, where owing to the confinement of quarks and gluons, there
is no distinguished mass scale in the renormalization procedure.

The situation is different in QED and in the electroweak theory. There
the classical Thomson scattering and the particle masses set natural scales
where the parameters can be defined. In QED the favoured renormalization
scheme is the on-shell scheme, where $e = \sqrt{4\pi\alpha}$ and the charged-fermion
masses are used as input parameters. The finite parts of the counterterms

are fixed by the renormalization conditions that the fermion propagators have poles at their physical masses, and that e becomes the $ee\gamma$ coupling constant in the Thomson limit of Compton scattering. In the electroweak Standard Model a distinguished set for parameter renormalization is given in terms of e, m_Z, m_W, m_H, m_f, with the masses of the corresponding particles. This electroweak on-shell scheme is the straightforward extension of the familiar QED renormalization, first proposed by Ross and Taylor [12] and used in many practical applications [8,13–22]. The mass of the Higgs boson, as long as it is experimentally unknown, is treated as a free input parameter. The light-quark masses can be considered only as effective parameters. In the cases of practical interest they can be replaced in terms of directly measured quantities like the cross-section for $e^+e^- \rightarrow$ hadrons.

The electroweak mixing angle is related to the vector boson masses in general by

$$\sin^2 \theta_W = 1 - \frac{m_W^2}{\rho_0\, m_Z^2}, \qquad (1.45)$$

where $\rho_0 \neq 1$ at the tree level in the case of a Higgs system more complicated than one with doublets only. We want to restrict our discussion of radiative corrections primarily to the minimal model with $\rho_0 = 1$. For $\rho_0 \neq 1$ see Sect. 1.5.

Instead of the set e, m_W, m_Z as basic free parameters one may alternatively use as basic parameters α, G_μ, m_Z [23] or $\alpha, G_\mu, \sin^2 \theta_W$ with the mixing angle deduced from neutrino–electron scattering [24] or perform the loop calculations in the $\overline{MS}$ scheme [25–28]. In the so-called $*$-scheme [29,30] a different method of bookkeeping in terms of effective running couplings was applied. Here we follow the line of the on-shell scheme as specified in detail in [8,21], but skip field renormalization.

Before predictions can be made from the theory, a set of independent parameters has to be taken from experiment. For practical calculations the physical input quantities $\alpha, G_\mu, m_Z, m_f, m_H, \alpha_s$ are commonly used to fix the free parameters of the Standard Model. The differences between various schemes are formally of higher order than the one under consideration. The study of the scheme dependence of the perturbative results, after improvement by resummation of the leading terms, allows us to estimate the missing higher-order contributions.

For a proper treatment of the charged-current vertex at the one-loop level, the matrix U_{CKM} has to be renormalized as well. As was shown in [31], where the renormalization procedure was extended to U_{CKM}, the resulting effects are completely negligible for the known light fermions. We therefore skip the renormalization of U_{CKM} in our discussion of radiative corrections.

1.2.2 Mass Renormalization

We have now to discuss the one-loop contributions to the on-shell parameters and their renormalization. Since the boson masses are part of the propagators we have to investigate the effects of the W and Z self-energies.

We restrict our discussion to the transverse parts $\sim g_{\mu\nu}$. In the electroweak theory, unlike QED, the longitudinal components $\sim q_\mu q_\nu$ of the vector boson propagators do not give zero results in physical matrix elements. But for light external fermions the contributions are suppressed by $(m_f/m_Z)^2$ and we are able to neglect them. Writing the self-energies as

$$\Sigma_{\mu\nu}^{W,Z} = g_{\mu\nu}\Sigma^{W,Z} + \cdots \tag{1.46}$$

with scalar functions $\Sigma^{W,Z}(q^2)$, we have for the one-loop propagators ($V = W, Z$)

$$\frac{-ig^{\mu\sigma}}{q^2 - m_V^2}\left(-i\,\Sigma_{\rho\sigma}^V\right)\frac{-ig^{\rho\nu}}{q^2 - m_V^2} = \frac{-ig^{\mu\nu}}{q^2 - m_V^2}\left(\frac{-\Sigma^V(q^2)}{q^2 - m_V^2}\right) \tag{1.47}$$

(the insertion of the factor $-i$ in the self energy is a convention). Besides the fermion loop contributions in the electroweak theory, there are also the non-Abelian gauge boson loops and loops involving the Higgs boson. The Higgs boson thus enters the four-fermion amplitudes as an experimentally unknown object at the level of radiative corrections, with its mass to be treated as an additional free parameter. In the graphical representation, the self-energies for the vector bosons denote the sum of all the diagrams with virtual fermions, vector bosons, and Higgs and ghost loops.

Resumming all self-energy terms yields a geometric progression for the dressed propagators:

$$\frac{-ig_{\mu\nu}}{q^2 - m_V^2}\left[1 + \left(\frac{-\Sigma^V}{q^2 - m_V^2}\right) + \left(\frac{-\Sigma^V}{q^2 - m_V^2}\right)^2 + \cdots\right]$$

$$= \frac{-ig_{\mu\nu}}{q^2 - m_V^2 + \Sigma^V(q^2)}. \tag{1.48}$$

The locations of the poles in the propagators are shifted by the self-energies. Consequently, the principal step in *mass renormalization* consists in a re-interpretation of the parameters: the masses in the Lagrangian cannot be the physical masses of the W and Z^0 but are the "bare masses" related to the physical masses m_W, m_Z by

$$m_W^{0\,2} = m_W^2 + \delta m_W^2\,,$$
$$m_Z^{0\,2} = m_Z^2 + \delta m_Z^2\,, \tag{1.49}$$

with counterterms of one-loop order. The propagators corresponding to this prescription are given by

$$\frac{-ig_{\mu\nu}}{q^2 - m_V^{0\,2} + \Sigma^V(q^2)} = \frac{-ig_{\mu\nu}}{q^2 - m_V^2 - \delta m_V^2 + \Sigma^V(q^2)} \qquad (1.50)$$

instead of (1.48). The renormalization conditions which ensure that $m_{W,Z}$ are the physical masses fix the mass counterterms to be

$$\delta m_W^2 = \mathrm{Re}\,\Sigma^W(m_W^2)\,,$$
$$\delta m_Z^2 = \mathrm{Re}\,\Sigma^Z(m_Z^2)\,. \qquad (1.51)$$

In this way, two of our input parameters and their counterterms have been defined.

1.2.3 Charge Renormalization

Our third input parameter is the electromagnetic charge e. The electroweak *charge renormalization* is very similar to that in pure QED. As in QED, we want to maintain the definition of e as the classical charge in the Thomson cross-section

$$\sigma_{\mathrm{Th}} = \frac{e^4}{6\pi\, m_e^2}\,. \qquad (1.52)$$

Accordingly, the Lagrangian carries the bare charge $e_0 = e + \delta e$ with the charge counterterm δe of one-loop order. The charge counterterm δe has to absorb the electroweak loop contributions to the $ee\gamma$ vertex in the Thomson limit. This charge renormalization condition is simplified by the validity of a generalization of the QED Ward identity [32] which implies that those corrections related to the external particles cancel each other. Thus for δe only two universal contributions are left (we use $s_W = \sin\theta_W$, $c_W = \cos\theta_W$ as abbreviations):

$$\frac{\delta e}{e} = \frac{1}{2}\,\Pi^\gamma(0) - \frac{s_W}{c_W}\,\frac{\Sigma^{\gamma Z}(0)}{m_Z^2}\,. \qquad (1.53)$$

The first contribution, quite in analogy to QED, is given by the photon vacuum polarization (see (1.55) for the definition of Π^γ) for real photons, $q^2 = 0$. But now, besides the fermion loops, it contains also bosonic loop diagrams from W^+W^- virtual states and the corresponding ghosts. The second term contains the mixing between photon and Z^0, in general described as a mixing propagator with $\Sigma^{\gamma Z}$ normalized as

$$\Delta^{\gamma Z} = \frac{-ig_{\mu\nu}}{q^2}\left(\frac{-\Sigma^{\gamma Z}(q^2)}{q^2 - m_Z^2}\right)\,. \qquad (1.54)$$

The fermion loop contributions to $\Sigma^{\gamma Z}$ vanish at $q^2 = 0$; only the non-Abelian bosonic loops yield $\Sigma^{\gamma Z}(0) \neq 0$.

To be more precise, the charge renormalization as discussed above is a condition only for the vector coupling constant of the photon. The axial coupling vanishes for on-shell photons as a consequence of the Ward identity.

From the diagonal photon self-energy

$$\Sigma^\gamma(q^2) = q^2\, \Pi^\gamma(q^2)\,, \tag{1.55}$$

no mass term arises for the photon since, in addition to the fermion loops, the bosonic loops also behave like

$$\Sigma^\gamma_{\text{bos}}(q^2) \simeq q^2\, \Pi^\gamma_{\text{bos}}(0) \to 0$$

for $q^2 \to 0$ leaving the pole at $q^2 = 0$ in the propagator. The absence of mass terms for the photon in all orders is a consequence of the unbroken electromagnetic gauge invariance.

In conclusion of this subsection, we can summarize the principal structure of electroweak one-loop calculations as follows:

- The classical Lagrangian $\mathcal{L}(e, m_W, m_Z, \ldots)$ is sufficent for lowest-order calculations and the parameters can be identified with the physical parameters.
- For higher-order calculations, $\mathcal{L}$ has to be considered as the "bare" Lagrangian of the theory $\mathcal{L}(e_0, m_W^0, m_Z^0, \ldots)$, with "bare" parameters which are related to the physical ones by

$$e_0 = e + \delta e, \quad m_W^{0\,2} = m_W^2 + \delta m_W^2, \quad m_Z^{0\,2} = m_Z^2 + \delta m_Z^2\,.$$

  The counterterms are fixed in terms of a certain subset of one-loop diagrams by specifying the definition of the physical parameters.
- For any four-fermion process we can write down the one-loop matrix element with the bare parameters and the loop diagrams for this process. Together with the counterterms the matrix element is finite when expressed in terms of the physical parameters, i.e. all UV singularities are removed.

1.2.4 Implications for Electroweak Parameters

We can now discuss the contributions to the electroweak parameter shifts which enter at the one-loop level via the renormalization procedure induced by the counterterms for the electric charge and for the electroweak mixing angle. Since these counterterms are universal, they appear everywhere where in the lowest-order expressions e or $\sin^2\theta_W$ is present. The shifts due to the counterterms are not finite. However, their finite parts contain large terms from the subclass of fermion loops.

$\Delta\alpha$ and Effective Electromagnetic Charge. Charge renormalization introduces the concept of electric charge for real photons ($q^2 = 0$), to be used for the calculation of observables at the electroweak scale set by m_Z. The charge counterterm in (1.53) contains the photon vacuum polarization at $q^2 = 0$. Since for electroweak processes the natural scale is given by the Z^0 mass, we split off the subtracted part evaluated at m_Z^2:

$$\Pi^\gamma(0) = -\operatorname{Re}\hat{\Pi}^\gamma(m_Z^2) + \operatorname{Re}\Pi^\gamma(m_Z^2)\,. \tag{1.56}$$

The UV-finite difference

$$\text{Re}\, \hat{\Pi}^\gamma(m_Z^2) = \text{Re}\, \Pi^\gamma(m_Z^2) - \Pi^\gamma(0) \tag{1.57}$$

of the photon vacuum polarization is a basic entry in the predictions for electroweak precision observables. The purely fermionic contributions correspond to standard QED and do not depend on the details of the electroweak theory. They are conveniently split into a leptonic and a hadronic contribution

$$\text{Re}\, \hat{\Pi}^\gamma(m_Z^2)_{\text{ferm}} = \text{Re}\, \hat{\Pi}^\gamma_{\text{lept}}(m_Z^2) + \text{Re}\, \hat{\Pi}^\gamma_{\text{had}}(m_Z^2) + \hat{\Pi}^\gamma_{\text{top}}(m_Z^2), \tag{1.58}$$

where the top quark is not included in the hadronic part from the five light flavours; it yields a small non-logarithmic contribution

$$\hat{\Pi}^\gamma_{\text{top}}(m_Z^2) \simeq \frac{\alpha}{\pi} Q_t^2 \frac{m_Z^2}{5\, m_t^2} \simeq 0.57 \times 10^{-4}. \tag{1.59}$$

The quantity

$$\begin{aligned} \Delta\alpha &= \Delta\alpha_{\text{lept}} + \Delta\alpha_{\text{had}} \\ &= -\text{Re}\, \hat{\Pi}^\gamma_{\text{lept}}(m_Z^2) - \text{Re}\, \hat{\Pi}^\gamma_{\text{had}}(m_Z^2) \end{aligned} \tag{1.60}$$

corresponds to a QED-induced shift in the electromagnetic fine structure constant

$$\alpha \rightarrow \alpha(1 + \Delta\alpha), \tag{1.61}$$

which can be resummed according to the renormalization group accommodating all the leading logarithms of the type $\alpha^n \log^n(m_Z/m_f)$. The evolution of the electromagnetic coupling with the scale μ is described by the renormalization group equation (RGE)

$$\mu \frac{d\alpha}{d\mu} = -\frac{\beta_0}{2\pi} \alpha^2, \tag{1.62}$$

where the coefficient of the one-loop β function in QED is

$$\beta_0 = -\frac{4}{3} \sum_{f \neq t} Q_f^2. \tag{1.63}$$

The solution of the RGE contains the leading logarithms in the resummed form. The result can be interpreted as an effective fine structure constant at the Z^0 mass scale:

$$\alpha(m_Z^2) = \frac{\alpha}{1 - \Delta\alpha}. \tag{1.64}$$

It corresponds to a resummation of the iterated one-loop vacuum polarization from the light fermions to all orders.

$\Delta\alpha$ is an input of crucial importance because of its universality and its remarkable size of $\sim 6\,\%$. The leptonic content can be directly evaluated in terms of the known lepton masses, yielding at one-loop order

$$\Delta\alpha_{\text{lept}} = \sum_{\ell=e,\mu,\tau} \frac{\alpha}{3\pi} \left(\log \frac{m_Z^2}{m_\ell^2} - \frac{5}{3} \right) + O\left(\frac{m_\ell^2}{m_Z^2} \right). \tag{1.65}$$

The two-loop correction has been known for a long time [33], and the three-loop contribution is now also available [34], yielding altogether

$$\Delta\alpha_{\text{lept}} = 314.97687 \times 10^{-4}$$
$$= (314.19007_{\text{1-loop}} + 0.77617_{\text{2-loop}} + 0.01063_{\text{3-loop}}) \times 10^{-4}. \tag{1.66}$$

For the light-hadronic part, perturbative QCD is not applicable and quark masses are not available as reasonable input parameters. Instead, the five-flavour contribution to $\hat{\Pi}_{\text{had}}^\gamma$ can be derived from experimental data with the help of the dispersion relation

$$\Delta\alpha_{\text{had}} = -\frac{\alpha}{3\pi} m_Z^2 \, \text{Re} \int_{4m_\pi^2}^{\infty} ds' \frac{R^\gamma(s')}{s'(s' - m_Z^2 - i\varepsilon)} \tag{1.67}$$

with

$$R^\gamma(s) = \frac{\sigma(e^+e^- \to \gamma^* \to \text{hadrons})}{\sigma(e^+e^- \to \gamma^* \to \mu^+\mu^-)}$$

as an experimental input quantity in the problematic low-energy range.

Integrating by means of the trapezoidal rule (averaging data in bins) over e^+e^- data for the energy range below 40 GeV and applying perturbative QCD for the high-energy region above 40 GeV, the expression (1.67) yields the value [35,36]

$$\Delta\alpha_{\text{had}} = 0.0280 \pm 0.0007, \tag{1.68}$$

which is compatible with another independent analysis [37] with a different error treatment. Because of the lack of precision in the experimental data a large uncertainty is associated with the value of $\Delta\alpha_{\text{had}}$, which propagates into the theoretical error of the predictions of electroweak precision observables. Including additional data from τ decays [38] yields about the same result with a slightly improved uncertainty. Recently other attempts have been made to increase the precision of $\Delta\alpha$ [39,41–43] by "theory-driven" analyses of the dispersion integral (1.67). The common basis is the application of perturbative QCD down to the energy scale given by the τ mass for the calculation of the quantity $R^\gamma(s)$ outside the resonances. Those calculations were made possible by the recent availability of the quark-mass-dependent $O(\alpha_s^2)$ QCD corrections [44] for the cross-section down to close to the thresholds for b and c production. (The first step in this direction was made in [45] in the massless approximation.) In order to pin down the error, two different strategies are in use: the application of the method developed in [42] for minimizing the impact of data from less reliable regions, used in [39], and the rescaling of data in the open-charm region of 3.7–5 GeV from PLUTO/DASP/MARKII, for the purpose of normalization to agree with perturbative QCD, used in [41]. The results obtained for $\Delta\alpha_{\text{had}}$ are very similar:

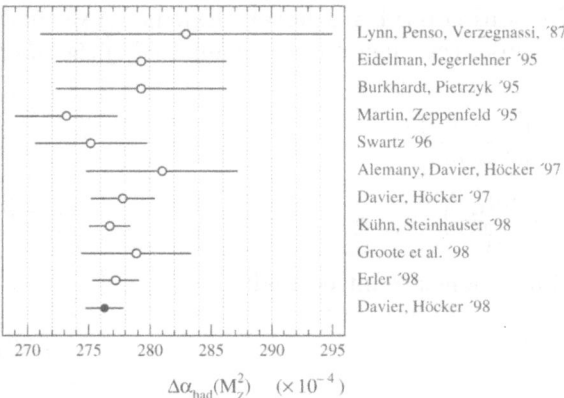

Lynn, Penso, Verzegnassi, '87
Eidelman, Jegerlehner '95
Burkhardt, Pietrzyk '95
Martin, Zeppenfeld '95
Swartz '96
Alemany, Davier, Höcker '97
Davier, Höcker '97
Kühn, Steinhauser '98
Groote et al. '98
Erler '98
Davier, Höcker '98

$$\Delta\alpha_{\text{had}}(M_Z^2) \quad (\times 10^{-4})$$

Fig. 1.1. Various determinations of $\Delta\alpha_{\text{had}}$ [40]

$$0.02763 \pm 0.00016 \quad [39], \tag{1.69}$$

$$0.02777 \pm 0.00017 \quad [41]. \tag{1.70}$$

In [43] the $\overline{MS}$ quantity $\hat{\alpha}(m_Z)$ was derived with the help of an unsubtracted dispersion relation in the $\overline{MS}$ scheme, yielding a comparable error. The history of the determination of the hadronic vacuum polarization is visualized in Fig. 1.1.

The basic assumption in the theory-driven approach, the validity of perturbative QCD and quark–hadron duality, is supported by the following empirical observations:

- The strong coupling constant $\alpha_{\text{s}}(m_\tau)$ determined from hadronic τ decays shows good agreement with $\alpha_{\text{s}}(m_Z)$ determined from Z^0-peak observables when the renormalization group evolution of α_{s} in perturbative QCD is imposed to run α_{s} from m_τ to the Z^0-mass scale.
- Non-perturbative contributions in $R^\gamma(s)$, parametrized in terms of condensates of quarks, gluons and vacuum expectation values of higher-dimensional operators in the operator product expansion [46], can be probed by comparing spectral moments of $R_{\text{exp}}^\gamma(s)$ with the corresponding expressions involving the theoretical R^γ. It has been shown from fitting a set of moments that the non-perturbative contributions are negligibly small [39,40].
- Recent preliminary measurements of R^γ at BES at 2.6 and 3.3 GeV show values slightly lower than the previous data [47], better in alignment with the expectations from perturbative QCD.

Although the error in the QCD-based evaluation of $\Delta\alpha_{\text{had}}$ is considerably reduced, the conservative estimate in (1.68) is independent of theoretical assumptions about QCD at lower energies and thus less sensitive to potential systematic effects [48] (see [49] for recent developments).

Renormalization of $\sin^2 \theta_W$ and the ρ Parameter. Expanding the bare mixing angle at one-loop order produces a counterterm composed of the mass counterterms:

$$s_W^{0\,2} = 1 - \frac{m_W^{0\,2}}{m_Z^{0\,2}} = 1 - \frac{m_W^2 + \delta m_W^2}{m_Z^2 + \delta m_Z^2}$$

$$= s_W^2 + c_W^2 \left(\frac{\delta m_Z^2}{m_Z^2} - \frac{\delta m_W^2}{m_W^2} \right). \tag{1.71}$$

Here the following shorthand notations are introduced:

$$c_W^2 = \frac{m_W^2}{m_Z^2}, \quad s_W^2 = 1 - c_W^2. \tag{1.72}$$

The explicit evaluation of the heavy-fermion part of the self-energies in (1.51) yields

$$\frac{\delta m_Z^2}{m_Z^2} = -N_C \frac{\alpha}{3\pi} \frac{3m_t^2}{8s_W^2 c_W^2 m_Z^2} \left(\Delta - \log \frac{m_t^2}{\mu^2} \right) + \cdots,$$

$$\frac{\delta m_W^2}{m_W^2} = -N_C \frac{\alpha}{3\pi} \frac{3m_t^2}{8s_W^2 m_W^2} \left(\Delta - \log \frac{m_t^2}{\mu^2} + \frac{1}{2} \right) + \cdots.$$

Here, the expression

$$\Delta = \frac{2}{\epsilon} - \gamma + \log 4\pi, \quad \epsilon = 4 - D, \tag{1.73}$$

and the arbitrary mass scale μ are the conventional UV parameters in dimensional regularization [50]. The difference, according to (1.71), displays a large shift in the mixing angle by a term $\sim m_t^2$, namely

$$s_W^{0\,2} = s_W^2 + c_W^2 \Delta \rho \tag{1.74}$$

with $(N_C = 3)$

$$\Delta \rho = N_C \frac{\alpha}{16\pi s_W^2 c_W^2} \frac{m_t^2}{m_Z^2}. \tag{1.75}$$

Since there are no other terms $\sim m_t^2$ in the mass counterterms $\delta m_Z^2, \delta m_W^2$ besides those which are q^2-independent, we have for this dominant part

$$\frac{\delta m_Z^2}{m_Z^2} - \frac{\delta m_W^2}{m_W^2} \simeq \frac{\Sigma^Z(0)}{m_Z^2} - \frac{\Sigma^W(0)}{m_W^2}, \tag{1.76}$$

which can be interpreted as a one-loop contribution to the ρ parameter

$$\rho = \frac{1}{1 - \Delta \rho}. \tag{1.77}$$

Originally the ρ parameter was defined as the ratio of the neutral- to the charged-current strength in neutrino scattering [51]. It contains the quantum correction

$$\Delta\rho = \frac{\Sigma^Z(0)}{m_Z^2} - \frac{\Sigma^W(0)}{m_W^2}, \tag{1.78}$$

which is finite when only heavy-fermion terms are considered. The term in (1.75) is the dominant contribution from the (t,b) doublet [52].

There is also a Higgs contribution to $\Delta\rho$ which, however, is not UV-finite by itself when derived from the diagrams involving the physical Higgs boson only. The m_H dependence for large Higgs masses m_H is only logarithmic [53]:

$$\Delta\rho_H \simeq \frac{g_2^2}{16\pi^2}\frac{3s_W^2}{4c_W^2}\log\frac{m_H^2}{m_W^2} + \cdots. \tag{1.79}$$

In contrast to the top term $\sim m_t^2$, the Higgs boson enters the shift of $\sin^2\theta_W$ in ways additional to that through $\Delta\rho$. There are additional m_H-dependent terms in $\delta m_Z^2, \delta m_W^2$ besides the ones in $\Sigma^W(0)$ and $\Sigma^Z(0)$. The same remark holds for the logarithmic top terms $\sim \log(m_t/m_W)$, which are not present in $\Delta\rho$.

The Correlation Between m_W and m_Z. Incorporating the parameter shifts

$$\alpha \to \alpha(1+\Delta\alpha), \quad s_W^2 \to s_W^2 + c_W^2\Delta\rho$$

into the tree-level relation (1.36), we obtain the approximate correlation at one-loop level

$$m_W^2\left(1 - \frac{m_W^2}{m_Z^2}\right) = \frac{\pi\alpha}{\sqrt{2}G_\mu}\left(1 + \Delta\alpha - \frac{c_W^2}{s_W^2}\Delta\rho + \cdots\right) \tag{1.80}$$

between the vector boson masses and the other electroweak parameters α and G_μ, taking into account the large contributions from light and heavy fermions. The ellipses indicate the residual terms belonging to the full calculation discussed in the next section.

The Neutral-Current Couplings. In a similar way to that above, we obtain a universal shift in the overall normalization of the neutral-current coupling constants in (1.25) and (1.27):

$$\frac{e}{2s_W c_W} \to \frac{e}{2s_W c_W}\left[1 + \frac{1}{2}\left(\Delta\alpha - \frac{c_W^2 - s_W^2}{s_W^2}\Delta\rho\right)\right]$$

$$= (\sqrt{2}G_\mu m_Z^2)^{1/2}\left(1 + \frac{\Delta\rho}{2}\right)$$

$$\to (\sqrt{2}G_\mu m_Z^2\rho_f)^{1/2}. \tag{1.81}$$

The complete expressions for the normalization factor

$$\rho_f = 1 + \Delta\rho + \cdots$$

and the effective mixing angle

$$s_f^2 = s_W^2 + c_W^2\Delta\rho + \cdots$$

in the Zff couplings between on-shell Z^0 bosons and fermions will be presented and discussed in the section on the Z^0 resonance.

1.3 The Vector Boson Mass Correlation

The interdependence between the gauge boson masses m_W, m_Z is established through the accurately measured muon lifetime or, equivalently, the Fermi coupling constant G_μ.

1.3.1 One-Loop Corrections to the Muon Lifetime

Originally, the μ lifetime τ_μ was calculated within the framework of the effective four-point Fermi interaction. Beyond the well-known one-loop QED corrections [54], the two-loop QED corrections in the Fermi model have been calculated quite recently [55], yielding the expression (the error in the two-loop term is from the hadronic uncertainty)

$$\frac{1}{\tau_\mu} = \frac{G_\mu^2 m_\mu^5}{192\pi^3} \left(1 - \frac{8m_e^2}{m_\mu^2}\right) \tag{1.82}$$

$$\times \left[1 + 1.810 \frac{\alpha}{4\pi} + (6.701 \pm 0.002)\left(\frac{\alpha}{4\pi}\right)^2\right].$$

This formula is the defining equation for G_μ in terms of the experimental μ lifetime. Owing to the presence of order-dependent QED corrections, the numerical value of the Fermi constant changes each time the next-order QED term becomes available. Including the second-order term, the most recent value is [55]

$$G_\mu = (1.16637 \pm 0.00001) \times 10^{-5}\,\text{GeV}^{-2}. \tag{1.83}$$

Within the Standard Model G_μ can be calculated in terms of the model parameters. In lowest order, the Fermi constant is given by the Standard Model expression (1.34) for the decay amplitude. In one-loop order, $G_\mu/\sqrt{2}$ coincides with the expression

$$\frac{G_\mu}{\sqrt{2}} = \frac{e_0^2}{8s_W^{0\,2}m_W^{0\,2}} \left[1 + \frac{\Sigma^W(0)}{m_W^2} + (VB)\right]. \tag{1.84}$$

This equation contains the bare parameters with the bare mixing angle. The term (VB) schematically summarizes the vertex corrections and box diagrams in the decay amplitude. A set of infrared-divergent "QED correction" graphs has been removed from this class of diagrams. These left-out diagrams, together with the real bremsstrahlung contributions, reproduce the QED correction factor of the Fermi model result in (1.82) and therefore have no influence on the relation between G_μ and the Standard Model parameters.

Equation (1.84) contains the bare parameters e_0, m_W^0, s_W^0. Expanding the bare parameters

$$e_0^2 = (e + \delta e)^2 = e^2 \left(1 + 2\frac{\delta e}{e}\right),$$

$$m_W^{0\,2} = m_W^2 \left(1 + \frac{\delta m_W^2}{m_W^2}\right),$$

$$s_W^{0\,2} = s_W^2 + c_W^2 \left(\frac{\delta m_Z^2}{m_Z^2} - \frac{\delta m_W^2}{m_W^2}\right) \tag{1.85}$$

and keeping only terms of one-loop order yields the expression

$$\frac{G_\mu}{\sqrt{2}} = \frac{e^2}{8 s_W^2 m_W^2}$$

$$\times \left[1 + 2\frac{\delta e}{e} - \frac{c_W^2}{s_W^2}\left(\frac{\delta m_Z^2}{m_Z^2} - \frac{\delta m_W^2}{m_W^2}\right) + \frac{\Sigma^W(0) - \delta m_W^2}{m_W^2} + (VB)\right]$$

$$\equiv \frac{e^2}{8 s_W^2 m_W^2}\,(1 + \Delta r)\,, \tag{1.86}$$

which is the one-loop corrected version of (1.34) [15].

The quantity $\Delta r(e, m_W, m_Z, m_H, m_t)$ is the finite combination of loop diagrams and counterterms in (1.86). Since we have already determined the counterterms in the previous subsection in terms of the boson self-energies, it is now only a technical problem to evaluate the one-loop diagrams for the final explicit expression of Δr. Here we quote the result:

$$(VB) = \frac{\alpha}{\pi s_W^2}\left(\Delta - \log\frac{m_W^2}{\mu^2}\right) + \frac{\alpha}{4\pi s_W^2}\left(6 + \frac{7 - 4s_W^2}{2s_W^2}\log c_W^2\right). \tag{1.87}$$

The singular part of this equation involving the divergence Δ, (1.73), coincides, up to a factor, with the non-Abelian bosonic contribution to the charge counterterm in (1.53):

$$\frac{\alpha}{\pi s_W^2}\left(\Delta - \log\frac{m_W^2}{\mu^2}\right) = \frac{2}{c_W s_W}\frac{\Sigma^{\gamma Z}(0)}{m_Z^2}\,. \tag{1.88}$$

Together with (1.53) and (1.87), we obtain the following from (1.86):

$$\Delta r = \Pi^\gamma(0) - \frac{c_W^2}{s_W^2}\left(\frac{\delta m_Z^2}{m_Z^2} - \frac{\delta m_W^2}{m_W^2}\right) + \frac{\Sigma^W(0) - \delta m_W^2}{m_W^2}$$

$$+2\frac{c_W}{s_W}\frac{\Sigma^{\gamma Z}(0)}{m_Z^2} + \frac{\alpha}{4\pi s_W^2}\left(6 + \frac{7 - 4s_W^2}{2s_W^2}\log c_W^2\right). \tag{1.89}$$

The first line is of particular interest: via $\Sigma^W(0)$ and the mass counterterms $\delta m_{W,Z}^2$, the masses m_H, m_t also enter Δr, whereas the residual terms depend only on the vector boson masses. We proceed with a more explicit discussion of the gauge-invariant subset of fermion loop corrections which involves, among others, the top quark. This subset is also of primary practical interest since it constitutes the numerically dominating part of Δr.

1.3.2 Fermion Contributions to Δr

The fermionic vacuum polarization of (1.89) is treated as described in Sect. 1.2.4:

$$\Pi^\gamma(0) = -\operatorname{Re}\hat{\Pi}^\gamma(m_Z^2) + \operatorname{Re}\Pi^\gamma(m_Z^2), \tag{1.90}$$

with the finite quantity $\hat{\Pi}^\gamma(m_Z^2)$ evaluated from (1.58).

Together with the unsubtracted quantity $\Pi^\gamma(m_Z^2)$, the W and Z self-energies enter the expression for Δr. For simplicity we restrict the further discussion to a single family, leptons or quarks, with $m_\pm$, $Q_\pm$, $v_\pm$, $a_\pm$ denoting the mass, charge, vector and axial-vector couplings of the up (+) and down (−) members. At the end, we perform the sum over the various families. We discuss the light and heavy fermions separately.

- Light fermions. In the light-fermion limit, i.e. neglecting all terms $\sim m_\pm/m_{W,Z}$, the various ingredients of Δr read (with an additional factor $N_{\mathrm{C}} = 3$ in the case of quark doublets)

$$\frac{\Sigma^W(0)}{m_W^2} = O\left(\frac{m_\pm^2}{m_W^2}\right) \simeq 0,$$

$$\frac{\delta m_Z^2}{m_Z^2} = \frac{\alpha}{3\pi}\frac{v_+^2 + a_+^2 + v_-^2 + a_-^2}{4s_W^2 c_W^2}\left(\Delta - \log\frac{m_Z^2}{\mu^2} + \frac{5}{3}\right),$$

$$\frac{\delta m_W^2}{m_W^2} = \frac{\alpha}{3\pi}\frac{1}{4s_W^2}\left(\Delta - \log\frac{m_Z^2}{\mu^2} + \frac{5}{3}\right) - \frac{\alpha}{16\pi s_W^2}\log c_W^2, \tag{1.91}$$

together with

$$\operatorname{Re}\Pi^\gamma(m_Z^2) = \frac{\alpha}{3\pi}(Q_+^2 + Q_-^2)\left(\Delta - \log\frac{m_Z^2}{\mu^2} + \frac{5}{3}\right). \tag{1.92}$$

Inserting everything into (1.89) yields

$$\Delta r = -\operatorname{Re}\hat{\Pi}^\gamma(m_Z^2)$$
$$+ \frac{\alpha}{3\pi}\left(\Delta - \log\frac{m_Z^2}{\mu^2} + \frac{5}{3}\right)$$
$$\times\left[Q_+^2 + Q_-^2 - \frac{c_W^2}{s_W^2}\left(\frac{v_+^2 + a_+^2 + v_-^2 + a_-^2}{4s_W^2 c_W^2} - \frac{1}{4s_W^2}\right) - \frac{1}{4s_W^4}\right]$$
$$- \frac{\alpha}{3\pi}\frac{c_W^2 - s_W^2}{4s_W^2}\log c_W^2$$

$$= -\operatorname{Re}\hat{\Pi}^\gamma(m_Z^2) - \frac{\alpha}{3\pi}\frac{c_W^2 - s_W^2}{4s_W^4}\log c_W^2. \tag{1.93}$$

The term in square brackets is zero with the coupling constants in (1.25). Thus, the main effect from the light fermions comes from the subtracted photon vacuum polarization, as the remnant from the renormalization of the electric charge at $q^2 = 0$.

- Heavy fermions. Of special interest is the case of the heavy top quark, which contributes a large correction $\sim m_t^2$ to Δr. In order to extract this piece we keep for simplicity only those terms which are either singular or quadratic in the top mass $m_t \equiv m_+$ ($N_C = 3$):

$$\mathrm{Re}\, \Pi^\gamma(m_Z^2) = N_C \frac{\alpha}{3\pi} (Q_+^2 + Q_-^2)\, \Delta_t,$$

$$\frac{\delta m_Z^2}{m_Z^2} = N_C \frac{\alpha}{3\pi} \left(\frac{v_+^2 + a_+^2 + v_-^2 + a_-^2}{4 s_W^2 c_W^2} - \frac{3 m_t^2}{8 s_W^2 c_W^2 m_Z^2} \right) \Delta_t,$$

$$\frac{\delta m_W^2}{m_W^2} = N_C \frac{\alpha}{3\pi} \left[\frac{1}{4 s_W^2} \Delta_t - \frac{3 m_t^2}{8 s_W^2 m_W^2} \left(\Delta_t + \frac{1}{2} \right) \right],$$

$$\frac{\Sigma^W(0)}{m_W^2} = -N_C \frac{\alpha}{3\pi} \frac{3 m_t^2}{8 s_W^2 m_W^2} \left(\Delta_t + \frac{1}{2} \right). \tag{1.94}$$

Here

$$\Delta_t = \Delta - \log \frac{m_t^2}{\mu^2}. \tag{1.95}$$

Inserting the above into (1.89), we verify that the singular parts cancel and a finite term $\sim m_t^2$ remains:

$$(\Delta r)_{b,t} = -\mathrm{Re}\hat{\Pi}_b^\gamma(m_Z^2) - \frac{c_W^2}{s_W^2} \Delta\rho + \cdots, \tag{1.96}$$

with the quantity $\Delta\rho$ from (1.75).

As a result of this discussion, we have recovered the same simple form for the leading terms as in (1.80), which remains also valid after including the full one-loop contribution:

$$\Delta r = \Delta\alpha - \frac{c_W^2}{s_W^2} \Delta\rho + (\Delta r)_{\mathrm{remainder}}. \tag{1.97}$$

$\Delta\alpha$ contains the large logarithmic corrections from the light fermions and $\Delta\rho$ the leading quadratic correction from a large top mass. All other terms are collected in $(\Delta r)_{\mathrm{remainder}}$. It should be noted that the remainder also contains a term logarithmic in the top mass (for which our approximation above was too crude), which is not negligible:

$$(\Delta r)_{\mathrm{remainder}}^{\mathrm{top}} = -\frac{\alpha}{4\pi s_W^2} \left(\frac{c_W^2}{s_W^2} - \frac{1}{3} \right) \log \frac{m_t}{m_Z} + \cdots. \tag{1.98}$$

Also, the Higgs boson contribution is part of the remainder. For large m_H, this contribution increases only logarithmically ("screening") [53]:

$$(\Delta r)_{\mathrm{remainder}}^{\mathrm{Higgs}} \simeq \frac{\alpha}{16\pi s_W^2} \frac{11}{3} \left(\log \frac{m_H^2}{m_W^2} - \frac{5}{6} \right). \tag{1.99}$$

The typical size of $(\Delta r)_{\mathrm{remainder}}$ is of the order ~ 0.01.

1.3.3 Higher-Order Contributions

Since Δr contains two large entries, $\Delta\alpha$ and $\Delta\rho$, a careful investigation of higher-order effects is necessary.

(i) Summation of $\Delta\alpha$ Terms. The replacement of the $\Delta\alpha$ part

$$1 + \Delta\alpha \ \rightarrow \ \frac{1}{1 - \Delta\alpha}$$

of the one-loop result in (1.86) correctly takes into account all orders in the leading logarithmic corrections $(\Delta\alpha)^n$, as shown by renormalization group arguments [56]. The evolution of the electromagnetic coupling with the scale μ is described by the renormalization group equation (1.62). The solution corresponds to a geometrical resummation of the vacuum polarization. Thus, in a situation where large corrections would be due only to the evolution of the electromagnetic charge between two very different scales set by m_f and m_Z, the resummed form

$$G_\mu = \frac{\pi\alpha}{\sqrt{2}m_W^2 s_W^2} \frac{1}{1 - \Delta r} = \frac{\pi\alpha}{\sqrt{2}m_Z^2 c_W^2 s_W^2} \frac{1}{1 - \Delta r} \qquad (1.100)$$

with Δr as in (1.97) would represent a good approximation to the full result.

(ii) Summation of $\Delta\rho$ Terms. The presence of the heavy top quark causes a large $\Delta\rho$ and the powers $(\Delta\rho)^n$ are not correctly resummed in (1.100). A result correct in the leading terms up to $O(\alpha^2)$ is instead given by the independent resummation [57]

$$\frac{1}{1 - \Delta r} \ \rightarrow \ \frac{1}{1 - \Delta\alpha} \frac{1}{1 + (c_W^2/s_W^2)\Delta\bar{\rho}} + (\Delta r)_{\text{remainder}} \,, \qquad (1.101)$$

where

$$\Delta\bar{\rho} = 3x_t[1 + x_t\rho^{(2)}(h)], \quad x_t = \frac{G_\mu m_t^2}{8\pi^2\sqrt{2}}, \quad h = \frac{m_H}{m_t}, \qquad (1.102)$$

incorporates the result from two-loop one-particle irreducible diagrams. For light Higgs bosons $m_H \ll m_t$, where m_H can be neglected, the coefficient

$$\rho^{(2)} = 19 - 2\pi^2 \qquad (1.103)$$

was first calculated in [58]. The general function $\rho^{(2)}$, valid for all Higgs masses, has been derived in [59]. With the resummed ρ parameter

$$\rho = \frac{1}{1 - \Delta\bar{\rho}} \,, \qquad (1.104)$$

(1.101) is compatible with the following form of the m_W–m_Z interdependence:

$$G_\mu = \frac{\pi}{\sqrt{2}} \frac{\alpha(m_Z^2)}{m_W^2 (1 - m_W^2/\rho m_Z^2)} [1 + (\Delta r)_{\text{remainder}}]$$

with

$$\alpha(m_Z^2) = \frac{\alpha}{1 - \Delta\alpha}. \tag{1.105}$$

It is interesting to compare this result with the corresponding lowest-order m_W–m_Z correlation in a more general model with a tree-level ρ parameter $\rho_0 \neq 1$: the tree-level ρ_0 enters in the same way as the ρ from a heavy top in the minimal model. The same applies for the quadratic mass terms from other particles such as scalars or additional heavy fermions in isodoublets with large mass splittings. Hence, up to the quantity $(\Delta r)_{\text{remainder}}$, the models are indistinguishable from an experimental point of view ($\Delta\alpha$ is universal). In the minimal model, however, ρ is calculable in terms of m_t, m_H, whereas ρ_0 is an *additional* free parameter.

(iii) QCD Corrections. Virtual gluons contribute to the quark loops in the vector boson self-energies at the two-loop level. For the light quarks this QCD correction is already contained in the result for the hadronic vacuum polarization from the dispersion integral (1.67). Fermion loops involving the top quark have additional $O(\alpha\alpha_s)$ corrections, which can be calculated perturbatively. The dominating term represents the QCD correction to the leading m_t^2 term of the ρ parameter and can be built in by writing, instead of (1.102),

$$\Delta\bar{\rho} = 3x_t \left(1 + x_t\, \rho^{(2)} + \delta\rho_{\text{QCD}}\right). \tag{1.106}$$

The QCD term [60,61] reads

$$\delta\rho_{\text{QCD}} = -\frac{\alpha_s(\mu)}{\pi} c_1 + \left(\frac{\alpha_s(\mu)}{\pi}\right)^2 c_2(\mu) \tag{1.107}$$

with

$$c_1 = \frac{2}{3}\left(\frac{\pi^2}{3} + 1\right) = 2.8599 \tag{1.108}$$

and the three-loop coefficient $c_2(\mu)$, which amounts to [61]

$$c_2 = -14.59 \quad \text{for } \mu = m_t \text{ and 6 flavours} \tag{1.109}$$

with the on-shell top mass m_t. This reduces the scale dependence of ρ significantly and hence is an important entry for decreasing the theoretical uncertainty of the Standard Model predictions for precision observables. As part of the higher-order irreducible contributions to ρ, the QCD correction is resummed together with the electroweak two-loop irreducible term as indicated in (1.101).

Beyond the $G_\mu m_t^2 \alpha_s$ approximation through the ρ parameter, the complete $O(\alpha\alpha_s)$ corrections to the self-energies are available from perturbative calculations [62] and by means of dispersion relations [63]. All the higher-order terms contribute with the same positive sign to Δr, making the dependence of Δr on the top mass significantly flatter. This is of high importance for the indirect determination of m_t from m_W measurements, which is affected by an amount of the order of 10 GeV. Also, non-leading terms in Δr of the type

$$\Delta r_{(bt)} = 3x_t \left(\frac{\alpha_s}{\pi}\right)^2 \left(a_1 \frac{m_Z^2}{m_t^2} + a_2 \frac{m_Z^4}{m_t^4}\right)$$

have been computed [64]. They contribute an extra term equal to $+0.0001$ to Δr; they are within the present uncertainty arising from $\Delta\alpha$.

(iv) Non-Leading Electroweak Higher-Order Terms. The modification of (1.101) by placing $(\Delta r)_{\text{remainder}}$ into the denominator

$$\frac{1}{1 - \Delta r} \rightarrow \frac{1}{(1 - \Delta\alpha)\left[1 + (c_W^2/s_W^2)\Delta\bar{\rho}\right] - (\Delta r)_{\text{remainder}}} \qquad (1.110)$$

correctly incorporates the non-leading higher-order terms containing mass singularities of the type $\alpha^2 \log(m_Z/m_f)$ [65].

In the meantime, the non-leading $G_\mu^2 m_t^2 m_Z^2$ contribution of the electroweak two-loop order in an expansion in terms of the top mass has been calculated [66]. This non-leading term turns out to be sizeable, about as large as the formally leading term of $O(m_t^4)$ obtained via the ρ parameter. In view of the present and future experimental accuracy it constitutes a non-negligible shift in the W mass.

Simultaneously, exact results have been derived for the Higgs dependence of the fermionic two-loop corrections in Δr [67], and comparisons have been performed with results obtained via the top mass expansion [68]. Differences in the values of m_W of several MeV (up to 8 MeV) are observed when m_H is varied over the range from 65 GeV to 1 TeV.

Figure 1.2 shows the dependence on the Higgs mass of the two-loop corrections to Δr associated with the t/b doublet, together with $\Delta\alpha$, the light-fermion terms not contained in $\Delta\alpha$ and the leading m_t^4-term, which constitutes a very poor approximation.

Pure fermion-loop contributions (n fermion loops at n-loop order) have also been investigated [68,69]. In the on-shell scheme, explicit results have been worked out up to four-loop order, which allows an investigation of the validity of the resummation (1.110) for the non-leading two-loop and higher-order terms. It was found that numerically the resummation (1.110) works remarkably well, within 2 MeV in m_W.

The correlation of the electroweak parameters, complete at the one-loop level and with the proper incorporation of the leading higher-order effects, is given by the following equation:

$$m_W^2 \left(1 - \frac{m_W^2}{m_Z^2}\right) = \frac{\pi\alpha}{\sqrt{2}G_\mu} \frac{1}{(1 - \Delta\alpha)\left[1 + (c_W^2/s_W^2)\Delta\bar{\rho}\right] - (\Delta r)_{\text{remainder}}}$$

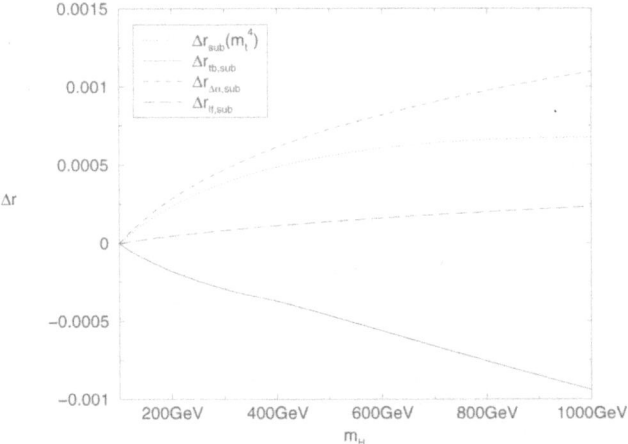

Fig. 1.2. Higgs mass dependence of fermionic contributions to Δr at the two-loop level [68]. The curves show the various contributions: light fermions via $\Delta\alpha$ ($\Delta r_{\Delta\alpha}$), residual light-fermion contribution not contained in $\Delta\alpha$ (Δr_{lf}), the contribution from the (tb) doublet (Δr_{tb}) and the approximation of the (tb) two-loop contribution by the term proportional to m_t^4. Displayed in each case is the difference $\Delta r(m_H) - \Delta r(100\ \mathrm{GeV})$

$$\equiv \frac{\pi\alpha}{\sqrt{2}G_\mu} \frac{1}{1-\Delta r} . \tag{1.111}$$

The term Δr in (1.111) is an effective quantity beyond the one-loop order, introduced to obtain a formal analogy to the naively resummed first-order result in (1.100). $\Delta\bar\rho$ includes the two-loop irreducible electroweak and QCD corrections to the ρ parameter according to (1.106). The correlation (1.111) allows us to predict a value for the W mass after the other parameters have been specified. The predicted value of m_W for $m_Z = 91.1867$ GeV, $m_t = 173.8 \pm 5.0$ GeV, $\alpha_s = 0.119$ and $\Delta\alpha_{\mathrm{had}}$ from (1.68) is given by

$$m_W = 80.300^{+0.074}_{-0.093} \pm 0.034 \ \mathrm{GeV} . \tag{1.112}$$

The central value is for a Higgs mass of 300 GeV; the upper and lower values belong to $m_H = 100$ and 1000 GeV; the second uncertainty corresponds to the experimental top mass error.

We can define the higher-order quantity Δr also as a physical observable by

$$\Delta r = 1 - \frac{\pi\alpha}{\sqrt{2}G_\mu} \frac{1}{m_W^2 \left(1 - m_W^2/m_Z^2\right)} . \tag{1.113}$$

Experimentally, this is determined by m_Z and m_W. Theoretically, it can be computed from m_Z, G_μ and α after specifying the masses m_H and m_t, by solving (1.111). Both electroweak and QCD higher-order effects yield a positive shift in Δr and thus diminish the slope of the first-order dependence

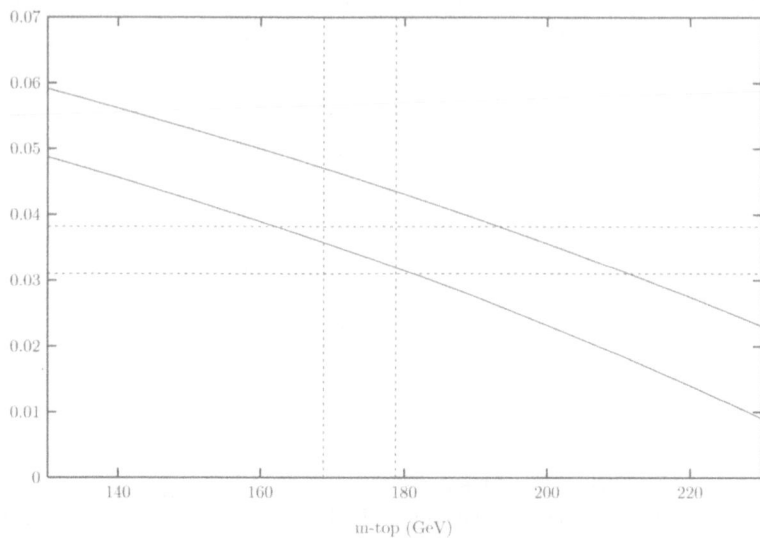

Fig. 1.3. Δr for $m_H = 70$ GeV (*lower solid line*) and $m_H = 1$ TeV (*upper solid line*), with experimental $\pm 1\sigma$ ranges for m_t and Δr

on m_t for large top masses. The effect on Δr from the Higgs dependent $\rho^{(2)}$ in (1.106) is an additional weakening of the sensitivity to m_t for large Higgs masses. The theoretical predictions of Δr for various Higgs and top masses is displayed in Fig. 1.3.

With the most recent data on m_Z and m_W, one obtains the experimental value

$$\Delta r = 0.0344 \pm 0.0027, \tag{1.114}$$

which is 12.7 standard deviations different from zero and 9 standard deviations away from the QED quantity $\Delta\alpha$ in (1.60). This clearly demonstrates not only the presence of quantum corrections but also in particular the genuine existence of the weak quantum effects of the Standard Model [70].

1.3.4 Relation to Deep-Inelastic Neutrino Scattering

The quantity s_W^2, or the ratio m_W/m_Z, can be measured indirectly in the class of low-energy experiments comprising neutrino–quark and neutrino–electron scattering. The two most precise pieces of informations come from the NC/CC neutrino–nucleon cross-section ratios [71,72]. For an isoscalar target these ratios do not depend on the nucleon structure [73]:

$$R_\nu = \frac{\sigma_{\mathrm{NC}}^\nu}{\sigma_{\mathrm{CC}}^\nu} = \left(\frac{m_W}{m_Z}\right)^4 \frac{1 - 2s_W^2 + (10/9)(1+r)\,s_W^4}{2(1-s_W^2)^2},$$

$$R_{\bar{\nu}} = \frac{\sigma^{\bar{\nu}}_{\text{NC}}}{\sigma^{\bar{\nu}}_{\text{CC}}} = \left(\frac{m_W}{m_Z}\right)^4 \frac{1 - 2s^2_W + (10/9)(1 + 1/r)\, s^4_W}{2(1 - s^2_W)^2}, \qquad (1.115)$$

with $r = \sigma^{\nu}_{\text{CC}}/\sigma^{\bar{\nu}}_{\text{CC}} \simeq 0.4$.

The second factor in R_ν has a very weak dependence on s^2_W. Hence, measurements of R_ν can be converted directly into values of m_W/m_Z. This principal feature remains valid also after the incorporation of radiative corrections in (1.115). In addition to the QED corrections, vertex corrections and box diagrams which do not depend on m_t and m_H, the dominant effect $\sim m^2_t$ can be simply embedded in (1.115) by means of the replacement

$$s^2_W \to s^2_W + c^2_W \Delta\rho,$$

according to (1.74). Since an increase in m_t is equivalent to a shift in s^2_W, the relation between R_ν and $(m_W/m_Z)^4$ is affected only marginally. This explains qualitatively the stability of m_W/m_Z against variations of m_t when extracted from R_ν.

1.4 Electroweak Physics at the Z^0 Resonance

The measurement of the Z^0 mass from the Z^0 lineshape at LEP provides us with an additional precise input parameter besides α and G_μ. Other observable quantities from the Z^0 peak, such as total and partial decay widths, asymmetries and τ-polarization, allow us to perform precision tests of the theory by comparison with the theoretical predictions.

1.4.1 Amplitudes and Effective Couplings

In lowest order, the processes $e^+e^- \to f\bar{f}$ are described by diagrams with photon and Z^0 exchange (Fig. 1.4). The one-loop diagrams with virtual-photon insertions (virtual QED corrections) form a gauge-invariant subset of the full one-loop terms. The sum of the virtual-photon loop graphs is UV-finite but IR-divergent because of the massless photon. The IR divergence is cancelled by adding the cross-section for real-photon bremsstrahlung (after integrating over the phase space for experimentally invisible photons), which always accompanies a realistic scattering process. Since the phase space for invisible photons is a detector-dependent quantity, the QED corrections cannot in general be separated from the details of the experimental device.

The residual set of one-loop diagrams, without any real or virtual photon, is commonly called the non-QED or weak corrections. This class is free of IR singularities but sensitive to details beyond the lowest-order amplitudes. The UV-singular terms associated with the loop diagrams are cancelled by the counterterms of Sect. 1.2, as a consequence of renormalizability. The one-loop amplitude for $e^+e^- \to f\bar{f}$ contains the sum of the individual contributions to

Fig. 1.4. Lowest-order photon and Z^0 exchange amplitudes for $e^+e^- \to f\bar{f}$

the self-energies and vertex corrections, including the external-fermion self-energies via wave function renormalization. The essential steps are: expressing the tree diagrams in terms of the bare parameters e_0, $m_Z^{0\,2}$, $s_W^{0\,2}$, expanding the bare quantities according to (1.49) and (1.85), and inserting the counterterms given by (1.51) and (1.53). The total amplitude around the Z^0 pole can be cast into a form close to the lowest-order amplitude,

$$A(e^+e^- \to f\bar{f}) = A_\gamma + A_Z + (\text{box}) \,,$$

as the sum of a dressed-photon and a dressed-Z^0 exchange amplitude plus the contribution from the box diagrams, which are numerically not significant around the peak (relative contribution $< 10^{-4}$). For theoretical consistency (gauge invariance) the latter have to be retained; for practical purposes they can be neglected in Z^0 physics. Resummation of the iterated self-energy insertions in the photon and Z propagators brings the finite Z^0 decay width into the denominator, and treats the higher-order leading terms in the proper way. Since the leading terms arise from fermion loops only, we do not have problems with gauge invariance; the bosonic loop terms have to be understood as expanded to strict one-loop order. Numerically, their resummation yields small differences, which give an impression of the possible size of the yet unknown next-order terms.

Dressed-Photon Amplitude. The dressed photon exchange amplitude, [1] with $s = (p_{e^+} + p_{e^-})^2$,

$$A_\gamma = \frac{e^2}{1 + \hat{\Pi}^\gamma(s)} \frac{Q_e Q_f}{s}$$
$$\times [(1 + F_V^{\gamma e})\gamma_\mu - F_A^{\gamma e}\gamma_\mu\gamma_5] \otimes \left[(1 + F_V^{\gamma f})\gamma^\mu - F_A^{\gamma f}\gamma^\mu\gamma_5\right] , \quad (1.116)$$

contains $\hat{\Pi}^\gamma$ as the γ self-energy subtracted at $s = 0$. Writing $\hat{\Pi}^\gamma$ in the denominator takes into account the resummation of the leading logarithms from the light fermions, given around the Z^0 by

$$\hat{\Pi}^\gamma_{\text{ferm}}(s) = -0.0593 - \frac{40\alpha}{18\pi} \log \frac{s}{m_Z^2} \pm 0.0007 + i\frac{\alpha}{3} \sum_{f \neq t} Q_f^2 N_C^f . \quad (1.117)$$

---

[1] The spinors for the external-fermion wave functions are dropped.

The form factors $F_{V,A}(s)$ arise from the vertex correction diagrams together with the external-fermion self-energies. They vanish for real photons: $F_{V,A}^{\gamma e,f}(0) = 0$. The typical size of the various corrections is (real parts):

$$\hat{\Pi}^{\gamma}(m_Z^2) \simeq -0.06\,,$$
$$F_V^{\gamma e}(m_Z^2) \simeq F_A^{\gamma e}(m_Z^2) \simeq 10^{-3}\,.$$

For the region around the Z^0 peak, the photon vertex form factors are negligibly small.

Dressed-Z^0 Amplitude and Effective Neutral-Current Couplings.
More important is the weak dressing of the Z^0 exchange amplitude. Without the box diagrams, the corrections factorize and we obtain a result quite close to the Born amplitude:

$$A_Z = \sqrt{2}G_\mu m_Z^2 (\rho_e \rho_f)^{1/2}$$
$$\times \frac{[\gamma_\mu(I_3^e - 2Q_e s_W^2 \kappa_e) - I_3^e \gamma_\mu \gamma_5] \otimes [\gamma^\mu(I_3^f - 2Q_f s_W^2 \kappa_f) - I_3^f \gamma^\mu \gamma_5]}{s - m_Z^2 + \mathrm{i}(s/m_Z)\Gamma_Z}\,.$$
(1.118)

The weak corrections appear in terms of the fermion-dependent form factors ρ and κ in the coupling constants and in the width in the denominator.

The s dependence of the imaginary part is due to the s dependence of $\mathrm{Im}\,\Sigma^Z$; the linearization is completely sufficient in the resonance region. We postpone the discussion of the Z^0 width for the moment and continue with the form factors.

The form factors ρ and κ in (1.118) have *universal* parts (i.e. independent of the fermion species) and *non-universal* parts which explicitly depend on the type of the external fermions. The universal parts arise from the counterterms and the boson self-energies, the non-universal parts from the vertex corrections and the fermion self-energies in the external lines:

$$\rho_{e,f} = 1 + (\Delta\rho)_{\mathrm{univ}} + (\Delta\rho)_{\mathrm{non\text{-}univ}}\,,$$
$$\kappa_{e,f} = 1 + (\Delta\kappa)_{\mathrm{univ}} + (\Delta\kappa)_{\mathrm{non\text{-}univ}}\,.$$
(1.119)

In their leading terms the universal contributions are given by

$$(\Delta\rho)_{\mathrm{univ}} = \Delta\rho + \cdots\,,$$
$$(\Delta\kappa)_{\mathrm{univ}} = \frac{c_W^2}{s_W^2}\Delta\rho + \cdots\,,$$

with $\Delta\rho$ obtained from (1.75). This structure can be easily understood from (1.71), (1.74) and (1.81). To incorporate the next-order leading terms one has to perform the substitutions .

$$\rho_{e,f} = 1 + \Delta\rho + \cdots \rightarrow \frac{1}{1 - \Delta\bar{\rho}} + \cdots\,,$$
$$\kappa_{e,f} = 1 + \frac{c_W^2}{s_W}\Delta\rho + \cdots \rightarrow 1 + \frac{c_W^2}{s_W}\Delta\bar{\rho} + \cdots\,,$$
(1.120)

with $\Delta\bar{\rho}$ obtained from (1.106).

The factorized Z amplitude allows us to define neutral-current (NC) vertices at the Z^0 resonance with effective coupling constants $g_{V,A}^f$, equivalently to the use of ρ_f, κ_f:

$$
\begin{aligned}
\Gamma_\mu^{NC} &= \left(\sqrt{2} G_\mu m_Z^2 \rho_f\right)^{1/2} \left[(I_3^f - 2 Q_f s_W^2 \kappa_f)\gamma_\mu - I_3^f \gamma_\mu \gamma_5\right] \\
&= \left(\sqrt{2} G_\mu m_Z^2\right)^{1/2} \left(g_V^f \gamma_\mu - g_A^f \gamma_\mu \gamma_5\right) .
\end{aligned} \tag{1.121}
$$

The complete expressions for the effective couplings read as follows:

$$
\begin{aligned}
g_V^f &= \left(v_f + 2 s_W c_W Q_f \frac{\hat{\Pi}^{\gamma Z}(m_Z^2)}{1 + \hat{\Pi}^\gamma(m_Z^2)} + F_V^{Zf}\right) \left(\frac{1 - \Delta r}{1 + \hat{\Pi}^Z(m_Z^2)}\right)^{1/2} , \\
g_A^f &= \left(a_f + F_A^{Zf}\right) \left(\frac{1 - \Delta r}{1 + \hat{\Pi}^Z(m_Z^2)}\right)^{1/2} .
\end{aligned} \tag{1.122}
$$

Besides Δr, the building blocks are the following finite combinations of two-point functions evaluated at $s = m_Z^2$:

$$
\begin{aligned}
\hat{\Pi}^Z(s) &= \frac{\text{Re}\, \Sigma^Z(s) - \delta m_Z^2}{s - m_Z^2} - \Pi^\gamma(0) \\
&\quad + \frac{c_W^2 - s_W^2}{s_W^2} \left(\frac{\delta m_Z^2}{m_Z^2} - \frac{\delta m_W^2}{m_W^2} - 2\frac{s_W}{c_W}\frac{\Sigma^{\gamma Z}(0)}{m_Z^2}\right) , \\
\hat{\Pi}^{\gamma Z}(s) &= \frac{\Sigma^{\gamma Z}(s) - \Sigma^{\gamma Z}(0)}{s} - \frac{c_W}{s_W}\left(\frac{\delta m_Z^2}{m_Z^2} - \frac{\delta m_W^2}{m_W^2}\right) + 2\frac{\Sigma^{\gamma Z}(0)}{m_Z^2}
\end{aligned}
$$

$$\tag{1.123}$$

and the finite form factors $F_{V,A}$ at $s = m_Z^2$ from the vertex corrections (including the external-fermion wave function renormalizations)

$$
\begin{aligned}
\Lambda_\mu^{(1\text{-loop})} &= \frac{e}{2 s_W c_W} \\
&\quad \times \left(\gamma_\mu F_V^{Zf}(s) - \gamma_\mu \gamma_5 F_A^{Zf}(s) + I_3^f \gamma_\mu(1 - \gamma_5)\frac{c_W}{s_W}\frac{\Sigma^{\gamma Z}(0)}{m_Z^2}\right) .
\end{aligned}
$$

For the explicit expressions for the self-energies and the vertex corrections see [21]. Owing to the imaginary parts of the self-energies and vertices, the form factors and the effective couplings, respectively, are complex quantities.

Effective Mixing Angles. We can define effective mixing angles for a given fermion species f according to

$$
s_f^2 = \sin^2 \theta_f = \frac{1}{4\,|Q_f|} \left(1 - \frac{\text{Re}\, g_V^f}{\text{Re}\, g_A^f}\right) , \tag{1.124}
$$

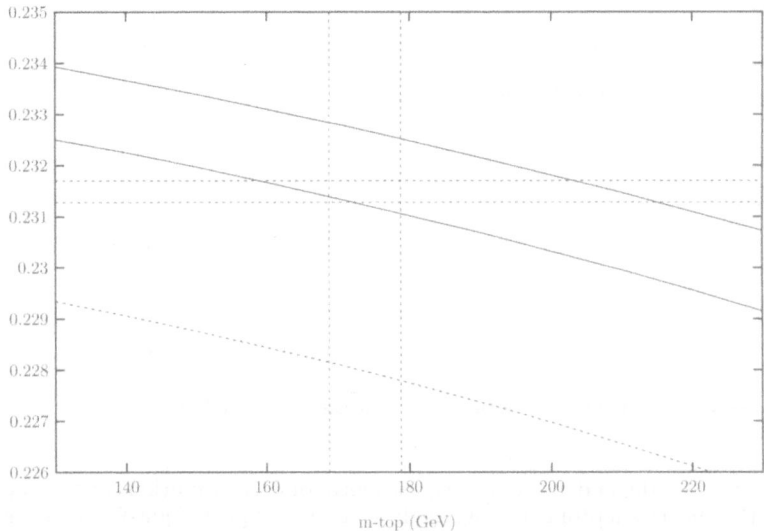

Fig. 1.5. $\sin^2 \theta_e$ for $m_H = 70$ GeV (*lower solid line*) and $m_H = 1$ TeV (*upper solid line*). Also shown are the experimental data ($\pm 1\sigma$ ranges) for m_t and s_e^2 (LEP/SLC average). The *dotted curve* includes only the fermionic loop effects via $\Delta\alpha$ and $\Delta\rho$

from the effective coupling constants in (1.122). They are of particular interest since they determine the on-resonance asymmetries, which will be discussed later in Sect. 1.4.4. Figure 1.5 displays the leptonic mixing angle s_e^2 and its compatibility with the data.

Beyond the leading two-loop corrections $\sim m_t^4$, which are related to the ρ parameter, the non-leading electroweak two-loop corrections $\sim G_\mu^2 m_t^2 m_Z^2$ in the top mass expansion for the leptonic mixing angle [66] s_ℓ^2 have become available, as well as those for ρ_ℓ [74].

Meanwhile, exact results have been derived for the Higgs dependence of the fermionic two-loop corrections in s_ℓ^2 [68,69], and comparisons were performed with those obtained via the top mass expansion [68]. The differences in the values of s_ℓ^2 can amount to 0.8×10^{-4} when m_H is varied over the range from 100 GeV to 1 TeV.

Figure 1.6 shows the Higgs-mass dependence of the 2-loop corrections to s_ℓ^2 associated with the t/b doublet, together with $\Delta\alpha$ and with the light-fermion terms not contained in $\Delta\alpha$. As can be seen, the m_H dependence of the light fermions yields contributions to s_ℓ^2 up to 2×10^{-5}. For ρ_ℓ or, equivalently, the leptonic Z^0 widths, the non-leading two-loop effects are small, and differences from the results for the top mass expansion are irrelevant.

The effective mixing angle is an observable that is very sensitive to the mass m_H of the Higgs boson. Since a light Higgs boson corresponds to a low value of s_ℓ^2, the most restrictive upper bound on m_H is from A_{LR} obtained at the SLC, which yields a rather low value for s_ℓ^2. The inclusion of the two-loop

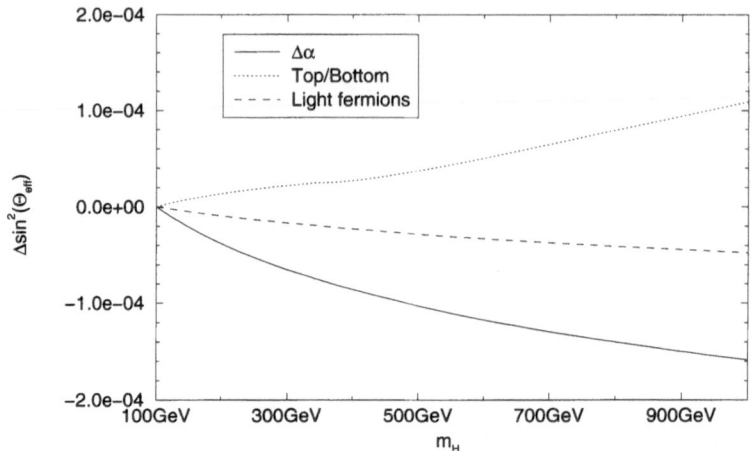

Fig. 1.6. Higgs-mass dependence of the various fermionic contributions at the two-loop level to the effective leptonic mixing angle s_ℓ^2 at the Z^0 peak: light-fermion contribution via $\Delta\alpha$, light-fermion contribution not contained in $\Delta\alpha$ and the contribution from the (tb) doublet. Shown in each case is the difference $s_\ell^2(m_H) - s_\ell^2(100\,\text{GeV})$

electroweak corrections $\sim m_t^2$ from [66] yields a sizeable positive contribution to s_ℓ^2 (see Fig. 1.7). The inclusion of this term hence strengthens the upper bound on m_H considerably.

The Zbb Couplings. The separation of a universal part in the effective couplings is sensible for two reasons: for the light fermions $(f \neq b, t)$ the non-universal contributions are small and (practically) independent of m_t and m_H, which enter only the universal part. This is, however, not true for the b quark, where the non-universal parts have a strong dependence on m_t [75] resulting from the virtual top quark in the vertex corrections. The difference between the d and b couplings can be parametrized in the following way

$$\rho_b = \rho_d(1+\tau)^2, \quad s_b^2 = s_d^2(1+\tau)^{-1}, \tag{1.125}$$

with the quantity

$$\tau = \Delta\tau^{(1)} + \Delta\tau^{(2)}$$

calculated perturbatively, comprising the complete one-loop-order term [75] with x_t from (1.102),

$$\Delta\tau^{(1)} = -2x_t - \frac{G_\mu m_Z^2}{6\pi^2\sqrt{2}}(c_W^2+1)\log\frac{m_t}{m_W} + \cdots, \tag{1.126}$$

and the leading electroweak two-loop contribution of $O(G_\mu^2 m_t^4)$ [59,76],

$$\Delta\tau^{(2)} = -2\,x_t^2\,\tau^{(2)}, \tag{1.127}$$

where $\tau^{(2)}$ is a function of m_H/m_t with $\tau^{(2)} = 9 - \pi^2/3$ for small m_H.

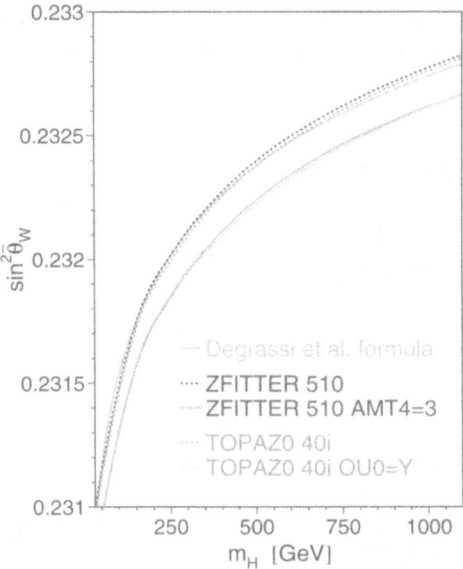

Fig. 1.7. Higgs-mass dependence of s_ℓ^2 with and without the electroweak two-loop term $\sim m_t^2$: comparison of ZFITTER and TOPAZ0 codes. The *lower* set of curves is obtained without, and the *upper* set with, the two-loop term (figure prepared by C. Pauss)

1.4.2 The Z^0 Lineshape

The integrated cross-section $\sigma(s)$ for $e^+e^- \to f\bar{f}$ around the Z^0 resonance with unpolarized beams is obtained from the formulae of the previous section in a straightforward way; it is expressed in terms of the effective vector and axial-vector coupling constants. It is, however, convenient to rewrite $\sigma(s)$ in terms of the Z^0 width and the partial widths Γ_e, Γ_f in order to have a more model-independent parametrization. The following form [77,78] includes final-state photon radiation and QCD corrections in the case of quark final states:[2]

$$\sigma(s) = \frac{12\pi\Gamma_e\Gamma_f}{\mid s - m_Z^2 + im_Z\Gamma_Z(s)\mid^2}\left(\frac{s}{m_Z^2} + R_f\frac{s - m_Z^2}{m_Z^2} + I_f\frac{\Gamma_Z}{m_Z} + \cdots\right)$$

$$+ \frac{4\pi\,\alpha(s)^2}{3s}\,Q_f^2 N_C^f\,K_{\mathrm{QCD}}\,(1 + \delta_{\mathrm{QED}}) \tag{1.128}$$

with

$$\Gamma_Z(s) = \Gamma_Z\left(\frac{s}{m_Z^2} + \epsilon\frac{s - m_Z^2}{m_Z^2} + \cdots\right) \tag{1.129}$$

---

[2] Since initial-state photon radiation is treated separately, the QED correction factor in (1.128) has to be removed in Γ_e.

and $N_C^f = 1$ for leptons and $N_C^f = 3$ for quarks. The terms R_f, I_f, ϵ are small quantities calculable in terms of the basic parameters. R_f and I_f describe the γ–Z interference,

$$R_f = \frac{2\,Q_e Q_f\,\mathrm{Re}(g_V^e g_V^f)}{(|g_V^e|^2 + |g_A^e|^2)(|g_V^f|^2 + |g_A^f|^2)}\,\frac{4\pi\alpha(s)}{\sqrt{2}G_\mu m_Z^2}\,,$$

$$I_f = \frac{2\,Q_e Q_f\,\mathrm{Re}(g_V^e g_V^f)}{(|g_V^e|^2 + |g_A^e|^2)(|g_V^f|^2 + |g_A^f|^2)}\,\frac{4\pi\alpha(s)}{\sqrt{2}G_\mu m_Z^2}\cdot\frac{s}{m_Z^2}\,\mathrm{Im}\,\hat{\Pi}^\gamma\,, \qquad (1.130)$$

and the last term is the QED background from pure photon exchange with

$$\alpha(s) = \frac{\alpha}{1 + \mathrm{Re}\,\hat{\Pi}_{\mathrm{ferm}}^\gamma(s)} \qquad (1.131)$$

and $\hat{\Pi}_{\mathrm{ferm}}^\gamma$ from (1.117). The small correction

$$\epsilon = \sum_f \epsilon_f\,, \quad \epsilon_f \simeq \frac{6m_f^2}{m_Z^2}\frac{\Gamma_f}{\Gamma_Z}\frac{|g_A^f|^2}{|g_V^f|^2 + |g_A^f|^2} \qquad (1.132)$$

is due to finite fermion-mass effects in the final states. I_f and ϵ have very little influence on the lineshape.

The s-dependent width gives rise to a displacement of the peak maximum by $\simeq -34$ MeV [79,80] The first term in the expansion (1.128) is the pure Z^0 resonance. It differs from a Breit–Wigner shape by the s dependence of the width:

$$\sigma_{\mathrm{res}}(s) = \sigma_0\frac{s\Gamma_Z^2}{(s - m_Z^2)^2 + s^2\Gamma_Z^2/m_Z^2}\,, \quad \sigma_0 = \frac{12\pi}{m_Z^2}\frac{\Gamma_e\Gamma_f}{\Gamma_Z^2}\,. \qquad (1.133)$$

By means of the substitution [80]

$$s - m_Z^2 + \mathrm{i}s\Gamma_Z/m_Z = (1 + \mathrm{i}\gamma)(s - \hat{M}_Z^2 + \mathrm{i}\hat{M}_Z\hat{\Gamma}_Z) \qquad (1.134)$$

with

$$\hat{M}_Z = m_Z(1 + \gamma^2)^{-1/2}\,, \quad \hat{\Gamma}_Z = \Gamma_Z(1 + \gamma^2)^{-1/2}\,, \quad \gamma = \frac{\Gamma_Z}{m_Z}\,, \qquad (1.135)$$

a Breit–Wigner resonance shape is recovered:

$$\sigma_{\mathrm{res}}(s) = \sigma_0\frac{s\hat{\Gamma}_Z^2}{(s - \hat{m}_Z^2)^2 + \hat{m}_Z^2\hat{\Gamma}_Z^2}\,. \qquad (1.136)$$

Numerically, one finds $\hat{M}_Z - m_Z \simeq -34$ MeV, $\hat{\Gamma}_Z - \Gamma_Z \simeq -1$ MeV. σ_0 is not changed. $\hat{M}_Z$ corresponds to the real part of the complex S-matrix pole s_0 of the Z^0 resonance [78,81], $s_0 = \hat{M}_Z^2 - \mathrm{i}\hat{M}_Z\hat{\Gamma}_Z$.

In the so-called S-matrix approach [82], the parameters of the resonance and those of the interference terms are considered as model-independent. Using the mass $\hat{M}_Z$ and width $\hat{\Gamma}_Z$ in the constant-width scheme, the total cross-section can be written as follows:

$$\sigma_{\text{tot}}(e^+e^- \to f\bar{f}) = \frac{4\pi\alpha^2}{3} \left[\frac{r_f^{\text{tot}}\, s + j_f^{\text{tot}}\,(s - \hat{M}_Z^2)}{(s - \hat{m}_Z^2)^2 + \hat{m}_Z^2 \hat{\Gamma}_Z^2} + \frac{g_f}{s} \right]. \qquad (1.137)$$

The second term in this sum corresponds to the pure photon exchange contribution where g_f can be taken over from QED as the product of the effective initial- and final-state charges. The four parameters $\hat{M}_Z, \hat{\Gamma}_Z, r_f^{\text{tot}}, j_f^{\text{tot}}$ are required to describe the energy dependence of the cross-section for a given fermion species in a more model-independent way; they can be derived from experimental data. In a similar way, in the same spirit of a model-independent parametrization, the antisymmetric part of the cross-section

$$\sigma_{\text{FB}} = \sigma_{\text{F}} - \sigma_{\text{B}},$$

with $\sigma_{\text{F,B}}$ defined in (1.163), can be expressed by two more quantities $r_f^{\text{fb}}, j_f^{\text{fb}}$ as follows:

$$\sigma_{\text{FB}}(e^+e^- \to f\bar{f}) = \frac{4\pi\alpha^2}{3} \left[\frac{r_f^{\text{fb}}\, s + j_f^{\text{fb}}\,(s - \hat{M}_Z^2)}{(s - \hat{m}_Z^2)^2 + \hat{m}_Z^2 \hat{\Gamma}_Z^2} \right]. \qquad (1.138)$$

QED Corrections. The observed cross-section is the result of convoluting (1.128) with the initial-state QED corrections, consisting of virtual-photon and real-photon bremsstrahlung contributions:

$$\sigma_{\text{obs}}(s) = \int_0^{x_{\max}} \mathrm{d}x\, H(x)\, \sigma[s(1-x)]. \qquad (1.139)$$

$x_{\max}$ denotes a cut on the radiated energy; $x = 2k/\sqrt{s}$. Kinematically $x_{\max}$ is limited to $1 - 4m_f^2/s$ or $1 - 4m_\pi^2/s$ for hadrons. For the required accuracy, multiple-photon radiation has to be included. The radiator function $H(x)$ with soft-photon resummation to all orders and the exact $O(\alpha^2)$ result [83,77], together with the leading logarithmic $O(\alpha^3)$ terms [84], has the following form:

$$H(x) = \beta x^{\beta-1} \delta^{\text{V}} + \delta^{\text{h}} \qquad (1.140)$$

with

$$\beta = \frac{2\alpha}{\pi}\,(L-1), \quad L = \log\frac{s}{m_e^2}; \qquad (1.141)$$

δ^{V} summarizes the virtual and soft-real-photon contributions, and δ^{h} the hard-photon contributions:

$$\delta^V = 1 + \frac{\alpha}{\pi}\delta_1^V + \left(\frac{\alpha}{\pi}\right)^2 \delta_2^V + \left(\frac{\alpha}{\pi}\right)^3 \delta_3^V,$$

$$\delta^h = \frac{\alpha}{\pi}\delta_1^h + \left(\frac{\alpha}{\pi}\right)^2 \delta_2^h + \left(\frac{\alpha}{\pi}\right)^3 \delta_3^h, \qquad (1.142)$$

with the following terms ($z = 1 - x$):

$$\delta_1^V = \frac{3}{2}L + \frac{\pi^2}{3} - 2,$$

$$\delta_2^V = \left(\frac{9}{8} - \frac{\pi^2}{3}\right)L^2 + \left(\frac{11}{12}\pi^2 + 3\zeta(3) - \frac{45}{16}\right)L$$

$$- \frac{\pi^4}{30} - \pi^2\log(2) + \frac{3\pi^2}{48} - \frac{9}{2}\zeta(3) + \frac{19}{4},$$

$$\delta_3^V = \left(\frac{9}{16} - \frac{\pi^2}{2} + \frac{8}{3}\zeta(3)\right)(L-1)^3; \qquad (1.143)$$

$$\delta_1^h = (L-1)(x-2),$$

$$\delta_2^h = (L-1)^2 \cdot 2(x-2)\log(x) + (L-1)\left(\frac{3}{2}L + \frac{\pi^2}{3} - 2\right)(x-2)$$

$$+ L^2\left[\left(\frac{1+z}{2} - \frac{1+z^2}{1-z}\right)\log(z) + z - 1\right]$$

$$+ L\left[\frac{1+z^2}{1-z}\left(\text{Li}_2(1-z) + \log(z)\log(x) + \frac{7}{2}\log(z) - \frac{1}{2}\log^2(z)\right)\right.$$

$$\left. + \frac{1+z}{4}\log^2(z) - \log(z) - 3z + \frac{7}{2}\right]$$

$$+ \frac{1+z^2}{1-z}X + (1+z)Y + Z,$$

$$\delta_3^h = (L-1)^3\frac{1}{6}\left[-\frac{27}{2} + \frac{15}{4}x + (4-2x)\left(\pi^2 - 6\log^2(x) + 3\text{Li}_2(x)\right)\right.$$

$$+ 3\left(-\frac{6}{x} + 7 - \frac{3}{2}x\right)\log(1-x) + \left(\frac{4}{x} - 7 + \frac{7}{2}x\right)\log^2(1-x)$$

$$\left. - 6(6-x)\log(x) + 6\left(-\frac{4}{x} + 6 - 3x\right)\log(x)\log(1-x)\right] \qquad (1.144)$$

with

$$X = -\frac{1}{6}\log^3(z) + \frac{1}{2}\log^2(z)\log(1-z)$$

$$- \log^2(z) - \frac{3}{2}\log(z)\log(1-z) + \left(\frac{\pi^2}{6} - \frac{17}{6}\right)\log(z)$$

$$+ \frac{1}{2}\log(z)\text{Li}_2(1-z) - \frac{3}{2}\text{Li}_2(1-z),$$

$$Y = \frac{3}{2}\text{Li}_3(1-z) - \log(1-z)\text{Li}_2(1-z) - 2S_{1,2}(1-z) - \frac{1}{2},$$

$$Z = -\frac{1-5z}{4}\log^2(1-z) + \frac{1-7z}{2}\log(z)\log(1-z) + \frac{11+10z}{6}\log(z)$$

$$+ \frac{3-2z}{2}\log(1-z) - \frac{25}{11}z\log^2(z) + \frac{2}{(1-z)^2}\log^2(z)$$

$$-\frac{2z}{3(1-z)}\left(1+\frac{\log(z)}{1-z}\right)^2-\frac{25}{6}z\,\mathrm{Li}_2(1-z)-\left(1-\frac{13}{3}z\right)\frac{\pi^2}{6},$$

where

$$\mathrm{Li}_2(u)=-\int_0^1 dt\,\frac{\log(1-ut)}{t}, \quad \mathrm{Li}_3(u)=\int_0^1 dt\,\frac{\log(t)\,\log(1-ut)}{t},$$

$$S_{1,2}(u)=\frac{1}{2}\int_0^1 dt\,\frac{\log^2(1-ut)}{t}.$$

The QED corrections have two major impacts on the line shape:

- a reduction of the peak height of the resonance cross-section by

$$\sigma_{\mathrm{obs}}^{\mathrm{peak}}\simeq\sigma_{\mathrm{res}}^{\mathrm{peak}}\left(\frac{\Gamma_Z}{m_Z}\right)^\beta\left(1+\frac{\alpha}{\pi}\delta_1^{\mathrm{V}}\right)\simeq 0.74\,\sigma_{\mathrm{res}}^{\mathrm{peak}};\tag{1.145}$$

- a shift in the peak position compared to the non-radiative cross-section by [77]

$$\Delta\sqrt{s_{\mathrm{max}}}=\frac{\beta\pi}{8}\Gamma_Z,\tag{1.146}$$

resulting in the following relation between the peak position and the nominal Z^0 mass:

$$\sqrt{s_{\mathrm{max}}}\simeq m_Z+\frac{\beta\pi}{8}\Gamma_Z-\frac{\Gamma_Z^2}{4m_Z}$$
$$\simeq m_Z+89\,\mathrm{MeV}.\tag{1.147}$$

It is important to note that, to high accuracy, these effects are practically universal, depending only on m_Z and Γ_Z as parameters. This allows a model-independent determination of these parameters from the measured lineshape.

A final remark concerns the QED corrections resulting from the interference between initial- and final-state radiation. These are not included in the treatment above, but they can be added in $O(\alpha)$ since they are small anyway. For cuts that are not too tight, as is the case for practical applications, these interference corrections to the lineshape are negligible and we do not list them here.

1.4.3 Z^0 Width and Partial Widths

From lineshape measurements one obtains the parameters m_Z, Γ_Z, σ_0, or the partial widths. Here m_Z will be used as a precise input parameter, together with α and G_μ; the width and partial widths are predictions of the Standard Model.

The total Z^0 width Γ_Z can be calculated as the sum over the partial decay widths

$$\Gamma_Z = \sum_f \Gamma_f, \quad \Gamma_f = \Gamma(Z \to f\bar{f}) \tag{1.148}$$

(other decay channels are not significant). The fermionic partial widths, when expressed in terms of the effective coupling constants defined in (1.122), read

$$\Gamma_f = \Gamma_0 \sqrt{1 - \frac{4m_f^2}{m_Z^2}} \left[|g_V^f|^2 \left(1 + \frac{2m_f^2}{m_Z^2}\right) + |g_A^f|^2 \left(1 - \frac{4m_f^2}{m_Z^2}\right) \right]$$
$$\times (1 + \delta_{\text{QED}}) + \Delta\Gamma_{\text{QCD}}^f \tag{1.149}$$

$$\simeq \Gamma_0 \left[|g_V^f|^2 + |g_A^f|^2 \left(1 - \frac{6m_f^2}{m_Z^2}\right) \right] (1 + \delta_{\text{QED}}) + \Delta\Gamma_{\text{QCD}}^f$$

with

$$\Gamma_0 = N_C^f \frac{\sqrt{2}G_\mu m_Z^3}{12\pi}. \tag{1.150}$$

The photonic QED correction, given at one-loop order by

$$\delta_{\text{QED}} = Q_f^2 \frac{3\alpha}{4\pi}, \tag{1.151}$$

is small; it's maximum is 0.17% for charged leptons.

The factorizable QCD corrections for hadronic final states can be written as follows:

$$\Delta\Gamma_{\text{QCD}}^f = \Gamma_0 \left(|g_V^f|^2 + |g_A^f|^2 \right) K_{\text{QCD}}, \tag{1.152}$$

where [85]

$$K_{\text{QCD}} = \frac{\alpha_s}{\pi} + 1.41 \left(\frac{\alpha_s}{\pi}\right)^2 - 12.8 \left(\frac{\alpha_s}{\pi}\right)^3 - \frac{Q_f^2}{4} \frac{\alpha\alpha_s}{\pi^2} \tag{1.153}$$

for the light quarks with $m_q \simeq 0$, with $\alpha_s = \alpha_s(m_Z^2)$.

For b quarks the QCD corrections are different owing to finite b mass terms and to top-quark-dependent two-loop diagrams for the axial part:

$$\Delta\Gamma_{\text{QCD}}^b = \Gamma_0 \left(|g_V^b|^2 + |g_A^b|^2 \right) K_{\text{QCD}}$$
$$+ \Gamma_0 \left[(g_V^b)^2 R_V + (g_A^b)^2 R_A \right]. \tag{1.154}$$

The coefficients in the perturbative expansions

$$R_V = c_1^V \frac{\alpha_s}{\pi} + c_2^V \left(\frac{\alpha_s}{\pi}\right)^2 + c_3^V \left(\frac{\alpha_s}{\pi}\right)^3 + \cdots,$$
$$R_A = c_1^A \frac{\alpha_s}{\pi} + c_2^A \left(\frac{\alpha_s}{\pi}\right)^2 + \cdots,$$

depending on m_b and m_t, have been calculated up to third order in α_s, except for the m_b-dependent singlet terms, which are known to $O(\alpha_s^2)$ [86,87]. For a review of the QCD corrections to the Z^0 width, with the explicit expressions for $R_{V,A}$, see [88].

The partial decay rate into b quarks, in particular the ratio $R_b = \Gamma_b/\Gamma_{\text{had}}$, is an observable with special sensitivity to the top quark mass. Therefore, beyond the pure QCD corrections, the two-loop contributions of the mixed QCD–electroweak type are also important. The QCD corrections were first derived for the leading term of $O(\alpha_s G_\mu m_t^2)$ [89] and were subsequently completed by the $O(\alpha_s)$ correction to the $\log m_t/m_W$ term [90] and the residual terms of $O(\alpha\alpha_s)$ [91].

At the same time, the complete two-loop $O(\alpha\alpha_s)$ corrections to the partial widths for decay into the light quarks have also been obtained, beyond those that are already contained in the factorized expression (1.152) with the electroweak one-loop couplings [92]. These "non-factorizable" corrections yield an extra negative contribution of $-0.55(3)$ MeV to the total hadronic Z^0 width. When converted into a shift of the strong coupling constant, they correspond to $\delta\alpha_s = 0.001$. In summary, the two-loop corrections of $O(\alpha\alpha_s)$ to the Z^0 widths are by now completely under control.

Radiation of secondary fermions through photons from the primary final-state fermions can yield another sizeable contribution to the partial Z^0 widths; however, this is compensated by the corresponding virtual contribution through the dressed-photon propagator in the final-state vertex correction for sufficiently inclusive final states, i.e. for loose cuts on the invariant mass of the secondary fermions [93].

In Table 1.1 the Standard Model predictions for the various partial widths and the total width of the Z^0 boson are collected. They include all the electroweak, QED and QCD corrections discussed above. Of particular interest are the following ratios of partial widths:

$$R_\ell = \frac{\Gamma_{\text{had}}}{\Gamma_\ell}, \quad R_b = \frac{\Gamma_b}{\Gamma_{\text{had}}}, \quad R_c = \frac{\Gamma_c}{\Gamma_{\text{had}}}. \tag{1.155}$$

Table 1.1. Partial and total Z^0 widths in MeV for $m_t = 175$ GeV and various Higgs masses (in GeV); $\alpha_s = 0.119$. Not listed are the values for $\Gamma_\tau = 0.9977\,\Gamma_e$ and Γ_c (very close to Γ_u)

m_H	Γ_ν	Γ_e	Γ_u	Γ_d	Γ_b	Γ_{had}	Γ_Z	R_ℓ
60	167.3	84.06	300.5	383.3	375.9	1743.5	2497.3	20.74
300	167.2	83.95	299.8	382.5	375.2	1740.0	2493.1	20.72
1000	166.9	83.81	299.0	381.7	374.2	1736.0	2488.0	20.71

1.4.4 Asymmetries

Left–Right Asymmetry. The left–right asymmetry is defined as the ratio

$$A_{\mathrm{LR}} = \frac{\sigma_{\mathrm{l}} - \sigma_{\mathrm{r}}}{\sigma_{\mathrm{l}} + \sigma_{\mathrm{r}}}, \tag{1.156}$$

where σ_{l} and σ_{r} denote the integrated cross-sections for left- and right-handed electrons, respectively. A_{LR}, in the case of lepton universality, is equal to the final-state polarization in τ pair production:

$$A_{\mathrm{pol}}^{\tau} = \mathcal{P}_{\tau} = A_{\mathrm{LR}} . \tag{1.157}$$

The on-resonance asymmetry ($s = m_Z^2$) in the improved Born approximation (real couplings $g_{\mathrm{V,A}}$) is given by

$$A_{\mathrm{LR}}(m_Z^2) = \mathcal{A}_e + \Delta A_{\mathrm{LR}}^{\mathrm{I}} + \Delta A_{\mathrm{LR}}^{Q} \tag{1.158}$$

where the combination

$$\mathcal{A}_e = \frac{2 g_{\mathrm{V}}^e g_{\mathrm{A}}^e}{(g_{\mathrm{V}}^e)^2 + (g_{\mathrm{A}}^e)^2} = \frac{2(1 - 4\sin^2\theta_e)}{1 + (1 - 4\sin^2\theta_e)^2} \tag{1.159}$$

depends only on the effective mixing angle (1.124) for the electron. The small contributions from the interference with the photon exchange

$$\Delta A_{\mathrm{LR}}^{\mathrm{I}} = \frac{2 Q_e Q_f \, g_{\mathrm{A}}^e g_{\mathrm{V}}^f}{(g_{\mathrm{V}}^{e\,2} + g_{\mathrm{A}}^{e\,2})(g_{\mathrm{V}}^{f\,2} + g_{\mathrm{A}}^{f\,2})} \frac{4\pi\alpha(m_Z^2)}{\sqrt{2} G_\mu m_Z^2} \frac{\Gamma_Z}{m_Z} \operatorname{Im}\hat{\Pi}^\gamma \tag{1.160}$$

and from the pure photon exchange part

$$\Delta A_{\mathrm{LR}}^{Q} = - \frac{\mathcal{A}_e \, Q_e^2 Q_f^2}{(g_{\mathrm{V}}^{e\,2} + g_{\mathrm{A}}^{e\,2})(g_{\mathrm{V}}^{f\,2} + g_{\mathrm{A}}^{f\,2})} \left(\frac{4\pi\alpha(m_Z^2)}{\sqrt{2} G_\mu m_Z^2}\right)^2 \left(\frac{\Gamma_Z}{m_Z}\right)^2 \tag{1.161}$$

are listed in Table 1.2 for the various final-state fermions. The contributions are negligibly small except for those from lepton final states. Mass effects from final fermions practically cancel. The same holds for QCD corrections in the case of quark final states, final-state QED corrections and QED corrections from the interference of initial- and final-state photon radiation. Initial-state QED corrections can be treated in complete analogy to (1.139), applied to $\sigma_{\mathrm{l,r}}(s)$. Their net effect on the asymmetry is also very small and practically independent of cuts [94,95]. A_{LR} thus represents a unique laboratory for testing the non-QED part of the electroweak theory. Measurements of A_{LR} are essentially measurements of $\sin^2\theta_e$ or of the ratio $g_{\mathrm{V}}^e/g_{\mathrm{A}}^e$.

Forward–Backward Asymmetries. The forward–backward asymmetry is defined by

$$A_{\mathrm{FB}} = \frac{\sigma_{\mathrm{F}} - \sigma_{\mathrm{B}}}{\sigma_{\mathrm{F}} + \sigma_{\mathrm{B}}} \tag{1.162}$$

with

Table 1.2. Contributions to the on-resonance left–right asymmetry for various final-state fermions; $\sin^2\theta_e = 0.2314$

f	A_e	$\Delta A_{\mathrm{LR}}^{\mathrm{I}}$	$\Delta A_{\mathrm{LR}}^{Q}$
μ	0.1511	0.0002	-0.0009
τ	0.1511	0.0002	-0.0009
c	0.1511	0.0005	-0.0003
b	0.1511	0.0004	-0.0001

$$\sigma_{\mathrm{F}} = \int_{\theta>\pi/2} \mathrm{d}\Omega\,\frac{\mathrm{d}\sigma}{\mathrm{d}\Omega}\,,\quad \sigma_{\mathrm{B}} = \int_{\theta<\pi/2} \mathrm{d}\Omega\,\frac{\mathrm{d}\sigma}{\mathrm{d}\Omega}\,. \tag{1.163}$$

For the on-resonance asymmetry ($s = m_Z^2$), we obtain in the improved Born approximation

$$A_{\mathrm{FB}}(m_Z^2) = A_{\mathrm{FB}}^{0,f}\left(1 - 4\mu_f + 6\mu_f\frac{(g_{\mathrm{A}}^f)^2}{(g_{\mathrm{V}}^f)^2 + (g_{\mathrm{A}}^f)^2}\right) + \Delta A_{\mathrm{FB}}^{\mathrm{I}} + \Delta A_{\mathrm{FB}}^{Q} \tag{1.164}$$

with

$$A_{\mathrm{FB}}^{0,f} = \frac{3}{4}\mathcal{A}_e\,\mathcal{A}_f \quad \text{and} \quad \mu_f = \frac{m_f^2}{m_Z^2}\,. \tag{1.165}$$

$\mathcal{A}_f$ is defined as

$$\mathcal{A}_f = \frac{2g_{\mathrm{V}}^f g_{\mathrm{A}}^f}{(g_{\mathrm{V}}^f)^2 + (g_{\mathrm{A}}^f)^2} = \frac{2(1 - 4\,|\,Q_f\,|\,s_f^2)}{1 + (1 - 4\,|\,Q_f\,|\,s_f^2)^2}\,, \tag{1.166}$$

using the shorthand notation for the effective mixing angle in (1.124). The small contributions $\Delta A_{\mathrm{FB}}^{\mathrm{I},Q}$ result from the the interference with the photon exchange

$$\Delta A_{\mathrm{FB}}^{\mathrm{I}} = \frac{3}{4}\frac{2Q_e Q_f\,g_{\mathrm{A}}^e g_{\mathrm{A}}^f}{(g_{\mathrm{V}}^{e\,2} + g_{\mathrm{A}}^{e\,2})\,(g_{\mathrm{V}}^{f\,2} + g_{\mathrm{A}}^{f\,2})}\frac{4\pi\alpha(m_Z^2)}{\sqrt{2}G_\mu m_Z^2}\frac{\Gamma_Z}{m_Z}\,\mathrm{Im}\,\hat{\Pi}^\gamma \tag{1.167}$$

and from the pure photon exchange part

$$\Delta A_{\mathrm{FB}}^{Q} = -\frac{3}{4}\frac{\mathcal{A}_e\mathcal{A}_f\,Q_e^2 Q_f^2}{(g_{\mathrm{V}}^{e\,2} + g_{\mathrm{A}}^{e\,2})\,(g_{\mathrm{V}}^{f\,2} + g_{\mathrm{A}}^{f\,2})}\left(\frac{4\pi\alpha(m_Z^2)}{\sqrt{2}G_\mu m_Z^2}\right)^2\left(\frac{\Gamma_Z}{m_Z}\right)^2. \tag{1.168}$$

The on-resonance asymmetries are essentially determined by the values of the effective mixing angles for e and f entering into the product $\mathcal{A}_e\mathcal{A}_f$.

Through $s_{e,f}^2$, the dependence of the asymmetries on the basic Standard Model parameters m_t, m_H is also fixed. The small corrections from finite-mass effects, from interference and from photon exchange can be considered as practically independent of the details of the model. For demonstration purpose we list in Table 1.3 the various terms in the on-resonance asymmetries according to (1.164) for a common value of the effective mixing angle $s_e^2 = s_f^2 = 0.2314$.

Table 1.3. On-resonance forward–backward asymmetries for $s_f^2 = 0.2314$

f	$A_{\mathrm{FB}}^{0,f}$	Mass correction	ΔA_{FB}^I	ΔA_{FB}^Q
μ	0.0171	$< 10^{-6}$	0.0018	-0.0001
τ	0.0171	1.3×10^{-5}	0.0018	-0.0001
c	0.0758	2.5×10^{-5}	0.0011	-0.0002
b	0.1061	1.5×10^{-5}	0.0004	-5×10^{-5}

A special role is played by the forward–backward asymmetries for b quarks, A_{FB}^b. The factor A_b in (1.164), owing accidentally to the assignment of the quantum numbers, is practically a constant in the Standard Model: $A_b = 0.935 \pm 0.001$. A_{FB}^b is therefore essentially determined by $\mathcal{A}_e$ and hence is an observable that is very sensitive to the electronic mixing angle via (1.159). A precise measurement of A_{FB}^b is thus a precise measurement of $\sin^2 \theta_e$.

QED Corrections. Initial-state photonic corrections have a strong impact on A_{FB} owing to its steep variation with energy. As in the case of the line-shape, the observed forward–backward asymmetry results from convoluting the s-dependent asymmetric cross-section $\sigma_{\mathrm{FB}}(s)$ with the photon spectrum $\tilde{H}(z)$ from initial-state radiation:

$$A_{\mathrm{FB}}(s) = \frac{1}{\sigma(s)} \int_{z_0}^1 \mathrm{d}z \, \frac{4z}{(1+z)^2} \, \tilde{H}(z) \, \sigma_{\mathrm{FB}}(zs), \quad z_0 \geq \frac{4m_f^2}{s}, \qquad (1.169)$$

with the integrated cross-section $\sigma(s)$ from (1.139). In terms of the effective complex couplings (1.122), the asymmetric cross-section reads

$$\sigma_{\mathrm{FB}}(s) = 2 Q_e Q_f \left(\sqrt{2} G_\mu m_Z^2\right) \mathrm{Re}\left(g_A^e g_A^f \, \chi(s) \frac{e^2}{1 + \hat{\Pi}_{\mathrm{ferm}}^\gamma(s)^*} \right)$$
$$+ 4 \, \mathrm{Re}(g_V^e g_A^{e*}) \, \mathrm{Re}(g_V^f g_A^{f*}) \left(\sqrt{2} G_\mu m_Z^2\right)^2 |\chi(s)|^2 \qquad (1.170)$$

with

$$\chi(s) = \frac{s}{s - m_Z^2 + \mathrm{i}(s/m_Z)\Gamma_Z} \,. \tag{1.171}$$

The radiator function $\tilde{H}(z)$ is different from the radiator H for the symmetric cross-section in (1.140) in the hard-photon terms. In the present status of the calculation, $\tilde{H}(z)$ contains the exact $O(\alpha)$ contribution [96], the $O(\alpha^2)$ contribution in the leading logarithmic approximation [97] and the resummation of soft photons to all orders [98]:

$$\tilde{H}(z) = H(x) + h(z) \tag{1.172}$$

with $H(x)$ from (1.140), $x = 1 - z$. The difference resulting from hard photons is given by

$$
h(z) = \left(\frac{\alpha}{\pi}\right)^2 L^2 \cdot \frac{1}{4} \left[\frac{(1-z)^3}{2z} - (1+z)\log(z) + 2(1-z) \right.
$$
$$
\left. + \frac{(1-z)^2}{\sqrt{z}} \left(\arctan\frac{1}{\sqrt{z}} - \arctan\sqrt{z} \right) \right]. \tag{1.173}
$$

Final-state corrections in the absence of cuts have only a small impact on the forward–backward asymmetry; they are at one-loop level given by [99,96]:

$$A_{\mathrm{FB}} \to A_{\mathrm{FB}} \left(1 - \frac{3\alpha(m_Z^2)}{4\pi} \right). \tag{1.174}$$

Unlike the situation for the integrated cross-section, initial–final-state interference also yields a contribution to the asymmetry, which, however, in the case of loose cuts, is very small. The application of more severe cuts, as might be necessary for the experimental analysis, increases both the final-state and the interference contributions and makes their proper treatment indispensable. These contributions are included in the program packages used for the data analysis and are not described here in detail. An overview can be found in [98].

Final-state QCD corrections, in the case of quark pair production, have a similar structure to the corresponding QED corrections at one-loop level:

$$A_{\mathrm{FB}} \to A_{\mathrm{FB}} \left(1 - \frac{\alpha_{\mathrm{s}}(m_Z^2)}{\pi} \right), \tag{1.175}$$

in the absence of cuts. Finite-mass effects have to be considered for b quarks only; they are discussed in [99]. Two-loop QCD corrections in the massless approximation have also become available [100].

1.4.5 Accuracy of the Standard Model Predictions

The FORTRAN codes ZFITTER [101] and TOPAZ0 [102] have been updated by incorporating all the recent precision calculation results that were discussed in the previous section. Comparisons have shown good agreement between the predictions from the two independent programs [103]. The global fits of the Standard Model parameters to the electroweak precision data performed by the LEP Electroweak Working Group are based on these recent versions.

For a discussion of the theoretical reliability of the Standard Model predictions, one has to consider the various sources contributing to their uncertainties.

Parametric uncertainties result from the limited precision in the experimental values of the input parameters, essentially $\alpha_s = 0.119 \pm 0.002$ [104], $m_t = 173.8 \pm 5.0$ GeV, $m_b = 4.7 \pm 0.2$ GeV and the hadronic vacuum polarization as discussed in Sect. 1.2.4 The uncertainty of $\Delta\alpha_{\mathrm{had}}$ induces an error

$$\frac{\delta m_W}{m_W} = \frac{s_W^2}{c_W^2 - s_W^2} \frac{\delta(\Delta\alpha_{\mathrm{had}})}{2(1 - \Delta r)}$$

in the W mass. The conservative estimate of the error given in (1.68) leads to $\delta m_W = 13$ MeV in the W mass prediction, and $\delta \sin^2 \theta = 0.00023$ common to all of the mixing angles.

The uncertainties from the QCD contributions can essentially be traced back to those in the top quark loops in the vector boson self-energies. The knowledge of the $O(\alpha_s^2)$ corrections to the ρ parameter and Δr yields a significant reduction; they are small, although not negligible (e.g. $\sim 3 \times 10^{-5}$ in s_ℓ^2).

The size of unknown higher-order contributions can be estimated by using different treatments of non-leading terms of higher order in the implementation of radiative corrections in electroweak observables ("options") and by investigations of the scheme dependence. Explicit comparisons between the results of five different computer codes based on on-shell and $\overline{MS}$ calculations for the Z^0-resonance observables are documented in the "Electroweak Working Group Report" [105]. The inclusion of the non-leading two-loop corrections $\sim G_\mu^2 m_t^2 m_Z^2$ reduces the uncertainty in m_W to about 4 MeV and that in s_ℓ^2 below 10^{-4}, typically to $\pm 4 \times 10^{-5}$ [103,106].

Another theoretical uncertainty is associated with the experimental determination of the e^+e^- luminosity. This uncertainty enters through the reference process of Bhabha scattering at small angles and is given by the theoretical uncertainty of the corresponding calculation. At small scattering angles θ, Bhabha scattering is essentially a QED process and the cross-section is dominated by the t channel exchange of a photon:

$$\frac{\mathrm{d}\sigma}{\mathrm{d}\theta} = \frac{\pi\alpha^2}{s} \frac{\sin\theta}{(1 - \cos\theta)^2} \left[4 + (1 + \cos\theta)^2\right] . \tag{1.176}$$

The cross-section is only very little influenced by the Z^0 exchange contributions and their interference with the photon, but it is sizeably affected by QED corrections, i.e. by higher-order diagrams with virtual-photon insertions and real-photon bremsstrahlung. For a review see e.g. [107]. Recent improvements in the calculation of the $O(\alpha^2)$ next-to-leading logarithmic contributions are an important step in pinning down the theoretical error to below 0.1% [147].

1.5 Beyond the Minimal Model

We shall conclude the theoretical discussion with an overview on renormalizable generalizations of the minimal model and their effect on electroweak observables. Extended models can be classified into the following three categories:

(i) extensions within the minimal gauge group $SU(2) \times U(1)$ with $\rho_{\text{tree}} = 1$;
(ii) extensions within $SU(2) \times U(1)$ with $\rho_{\text{tree}} \neq 1$;
(iii) extensions with larger gauge groups $SU(2) \times U(1) \times G$ and their respective extra gauge bosons.

Extensions of the class (i) include, for example, models with additional (sequential) fermion doublets, models with more Higgs doublets and the minimal supersymmetric version of the Standard Model.

1.5.1 Generalization of Self-Energy Corrections

If "new physics" is present in the form of new particles which couple to the gauge bosons but not directly to the external fermions in a four-fermion process, only the self-energies are affected. In order to have a description which is as far as possible independent of the type of extra heavy particles, it is convenient to introduce a parametrization of the radiative corrections from the vector boson self-energies in terms of the static ρ parameter (here we use the notation $\Sigma^{ZZ} = \Sigma^Z$, $\Sigma^{WW} = \Sigma^W$)

$$\Delta\rho(0) = \frac{\Sigma^{ZZ}(0)}{m_Z^2} - \frac{\Sigma^{WW}(0)}{m_W^2} - 2\frac{s_W}{c_W}\frac{\Sigma^{\gamma Z}(0)}{m_Z^2} \tag{1.177}$$

and the combinations

$$\Delta_1 = \frac{1}{s_W}\Pi^{3\gamma}(m_Z^2) - \Pi^{33}(m_Z^2)\,,$$

$$\Delta_2 = \Pi^{33}(m_Z^2) - \Pi^{WW}(m_W^2)\,,$$

$$\Delta\alpha = \Pi^{\gamma\gamma}(0) - \Pi^{\gamma\gamma}(m_Z^2)\,. \tag{1.178}$$

The quantities in (1.178) are the isospin components of the self-energies

$$\Sigma^{\gamma Z} = -\frac{1}{c_W}\left(\Sigma^{3\gamma} - s_W\Sigma^{\gamma\gamma}\right)\,,$$

$$\Sigma^{ZZ} = \frac{1}{c_W^2}\left(\Sigma^{33} - 2s_W\Sigma^{3\gamma} + s_W^2\Sigma^{\gamma\gamma}\right) \tag{1.179}$$

in the expansions

$$\text{Re}\,\Sigma^{ij}(k^2) = \Sigma^{ij}(0) + k^2 \Pi^{ij}(k^2)\,. \tag{1.180}$$

The Δ notation above was introduced in [108]. Several other conventions are used in the literature, for example

• the S, T, U parameters of [109] are related to (1.177) and (1.178) by

$$S = \frac{4s_W^2}{\alpha}\Delta_1, \quad T = \frac{1}{\alpha}\Delta\rho(0), \quad U = \frac{4s_W^2}{\alpha}\Delta_2; \tag{1.181}$$

• the ϵ parameters of [110] are related by

$$\epsilon_1 = \Delta\rho, \quad \epsilon_2 = -\Delta_2, \quad \epsilon_3 = \Delta_1\,. \tag{1.182}$$

Further literature can be found in [111–114]. The combinations (1.177) and (1.178) of self-energies contribute in a universal way to the electroweak parameters (the residual corrections not arising from self-energies are dropped since they are identical to the Standard Model ones):

• the m_W–m_Z correlation in terms of Δr:

$$\Delta r = \Delta\alpha - \frac{c_W^2}{s_W^2}\Delta\rho(0) - \frac{c_W^2 - s_W^2}{s_W^2}\Delta_2 + 2\Delta_1\,; \tag{1.183}$$

• the normalization of the NC couplings at m_Z^2

$$\Delta\rho_f = \Delta\rho(0) + \Delta_Z\,, \tag{1.184}$$

where the extra quantity

$$\Delta_Z = m_Z^2 \frac{d\,\Pi^{ZZ}}{dk^2}(m_Z^2) \tag{1.185}$$

in (1.185) is from the residue of the Z propagator at the peak; heavy particles decouple from Δ_Z;
• the effective mixing angles

$$s_f^2 = (1 + \Delta\kappa')\,\tilde{s}^2, \quad \tilde{s}^2 = \frac{1}{2}\left(1 - \sqrt{1 - \frac{4\pi\alpha(m_Z^2)}{\sqrt{2}G_\mu m_Z^2}}\right), \tag{1.186}$$

with

$$\Delta\kappa' = -\frac{c_W^2}{c_W^2 - s_W^2}\Delta\rho(0) + \frac{\Delta_1}{c_W^2 - s_W^2}\,. \tag{1.187}$$

The finite combinations of self-energies (1.177) and (1.178) are of practical interest since they can be extracted from precision data in a fairly model-independent way. An experimental observable particularly sensitive to Δ_1 is the weak charge Q_W, which determines the atomic parity violation in Caesium [112]

$$Q_W = -73.20 \pm 0.13 - 0.82\Delta\rho(0) - 102\Delta_1\,; \tag{1.188}$$

this is almost independent of $\Delta\rho(0)$.

The theoretical interest in the Δ quantities is based on their selective sensitivity to different kinds of new physics.

- $\Delta\alpha$ has contributions from light charged particles only, whereas heavy objects decouple.
- $\Delta\rho(0)$ is a measure of the violation of the custodial $SU(2)$ symmetry. It is sensitive to particles with large mass splittings in multiplets. As an example, a general doublet of fermions with masses m_1, m_2 causes a shift of ρ by [52]

$$\Delta\rho(0) = N_C \frac{G_\mu}{8\pi^2\sqrt{2}} \left(m_1^2 + m_2^2 - \frac{2m_1^2 m_2^2}{m_1^2 - m_2^2} \log\frac{m_1^2}{m_2^2} \right). \tag{1.189}$$

Another example is the Higgs bosons of a two-Higgs doublet model [117–120] with masses m_{H^+}, m_h, m_H, m_A and mixing angles β, α for the charged $H^\pm$ and the neutral h^0, H^0, A^0 Higgs bosons, yielding

$$\Delta\rho(0) = \frac{G_\mu}{8\pi^2\sqrt{2}} \left[\sin^2(\alpha - \beta) F(m_{H^+}^2, m_A^2, m_H^2) \right.$$
$$\left. + \cos^2(\alpha - \beta) F(m_{H^+}^2, m_A^2, m_h^2) \right] \tag{1.190}$$

with

$$F(x, y, z) = x + \frac{yz}{y - z}\log\frac{y}{z} - \frac{xy}{x - y}\log\frac{x}{y} - \frac{xz}{x - z}\log\frac{x}{z}.$$

For $m_{H^+} \gg m_{\text{neutral}}$ or vice versa one finds a positive contribution

$$\Delta\rho(0) \simeq \frac{G_\mu m_{H^+}^2}{8\pi^2\sqrt{2}} \quad \text{or} \quad \frac{G_\mu m_{\text{neutral}}^2}{8\pi^2\sqrt{2}} > 0. \tag{1.191}$$

Also, a negative contribution

$$\Delta\rho(0) < 0 \quad \text{for} \quad m_{h,H} < m_{H^+} < m_A \quad \text{and} \quad m_A < m_{H^+} < m_{h,H}$$

is possible in the unconstrained two-doublet model.

- Δ_1 is sensitive to chiral symmetry breaking by masses. In particular, a doublet of mass-degenerate heavy fermions yields a contribution

$$\Delta_1 = N_C^f \frac{G_\mu m_W^2}{12\pi^2\sqrt{2}}, \tag{1.192}$$

whereas the contribution of degenerate heavy fermions to $\Delta\rho(0)$ is zero. Hence, Δ_1 can directly count the number N_{deg} of mass degenerate fermion doublets:

$$\Delta_1^f = 4.5 \times 10^{-4} N_{\text{deg}}.$$

Δ_1 also has sizeable contributions in models with a large number of additional fermions, as in technicolor models. For example, $\Delta_1 \simeq 0.017$ for $N_{\text{TC}} = 4$ and one family of technifermions [109,115].

- A further quantity ϵ_b has been introduced [116] in order to parametrize specific non-universal left-handed contributions to the $Zb\bar{b}$ vertex via

$$g_A^b = g_A^d (1 + \epsilon_b), \quad g_V^b/g_A^b = \left(1 - \frac{4}{3}s_d^2 + \epsilon_b\right)(1 + \epsilon_b)^{-1}. \tag{1.193}$$

Phenomenologically, the ϵ_i are parameters which can be determined experimentally from the electroweak precision data.

1.5.2 Models with $\rho_{\text{tree}} \neq 1$

One of the basic relations of the minimal Standard Model is the tree-level correlation between the vector boson masses and the electroweak mixing angle

$$\rho_{\text{tree}} = \frac{m_W^2}{m_Z^2 \cos^2 \theta_W} = 1 .$$

The formulation of the electroweak theory in terms of a local gauge theory requires at least a single scalar doublet for breaking the electroweak symmetry $\text{SU}(2) \times \text{U}(1) \to \text{U}(1)_{\text{em}}$. In contrast to the fermion and vector boson part, very little is known empirically about the scalar sector. Without the assumption of minimality, quite a lot of options are at our disposal, including more complicated multiplets of Higgs fields. The presence of at least a triplet of Higgs fields gives rise to $\rho_{\text{tree}} = \rho_0 \neq 1$. As a consequence, the tree-level relations between the electroweak parameters have to be generalized according to

$$\sin^2 \theta_W \to s_\theta^2 = 1 - \frac{m_W^2}{\rho_0 m_Z^2} \tag{1.194}$$

and

$$\frac{G_\mu}{\sqrt{2}} = \frac{e^2}{8 s_\theta^2 m_W^2} = \frac{e^2}{8 s_\theta^2 c_\theta^2 \rho_0 m_Z^2} . \tag{1.195}$$

A complete discussion of radiative corrections requires the calculation of the extra loop diagrams from the non-standard Higgs sector and an extension of the renormalization procedure [10,121,122]. Since m_W, m_Z and $\sin^2 \theta_W$ (or ρ_0, equivalently) are now independent parameters, one extra renormalization condition is required. A natural condition would be to define the mixing angle for electrons $s_\theta^2 = s_e^2$ in terms of the ratio of the dressed coupling constants at the Z^0 peak

$$\frac{g_V^e}{g_A^e} = 1 - 4 s_e^2 ,$$

which is measurable in terms of the left–right or the forward–backward asymmetries. This fixes the counterterm for s_e^2 by means of

$$\frac{\delta s_e^2}{s_e^2} = \frac{c_e}{s_e} \frac{\text{Re}\, \Sigma^{\gamma Z}(m_Z^2)}{m_Z^2} + \frac{c_e}{s_e} \frac{\Sigma^{\gamma Z}(0)}{m_Z^2} + \Delta\kappa_e , \tag{1.196}$$

where $\Delta\kappa_e$ is the finite part of the electron–Z vertex correction. The counterterms for the other parameters α, m_Z are treated as usual. With this input, we obtain a renormalized ρ parameter and the corresponding counterterm for the bare ρ parameter $\rho_0 = \rho + \delta\rho$ as follows:

$$\rho = \frac{m_W^2}{m_Z^2 c_e^2},$$

$$\frac{\delta\rho}{\rho} = \frac{\delta m_W^2}{m_W^2} - \frac{\delta m_Z^2}{m_Z^2} + \frac{\delta s_e^2}{c_e^2}. \tag{1.197}$$

Other derived quantities are:

- the relation between m_W, G_μ, s_e^2:

$$m_W^2 = \frac{\pi\alpha}{\sqrt{2}G_\mu s_e^2}\frac{1}{1-\Delta r} \tag{1.198}$$

with

$$\Delta r = \frac{\Sigma^W(0) - \delta m_W^2}{m_W^2} + \Pi^\gamma(0) - \frac{\delta s_e^2}{s_e^2} + 2\frac{c_e}{s_e}\frac{\Sigma^{\gamma Z}(0)}{m_Z^2} + \delta_{\rm VB}, \tag{1.199}$$

where the vertex–box contribution is given by

$$\delta_{\rm VB} = \frac{\alpha}{4\pi s_e^2}\left(6 + \frac{10 - 10s_e^2 - 3(1-2s_e^2)c_W^2/c_e^2}{2(1-c_W^2)}\log c_W^2\right), \quad c_W = \frac{m_W}{m_Z};$$

- the normalization of the Zff couplings at one-loop level:

$$\frac{e^2}{4s_e^2 c_e^2}\left(1 + \Pi^\gamma(0) - \frac{c_e^2 - s_e^2}{c_e^2}\frac{\delta s_e^2}{s_e^2} + 2\frac{c_e^2 - s_e^2}{c_e s_e}\frac{\Sigma^{\gamma Z}(0)}{m_Z^2} + \Delta\rho_f\right)$$

$$= \sqrt{2}G_\mu m_Z^2\rho\left(1 - \frac{\Sigma^W(0) - \delta m_W^2}{m_W^2} + \frac{\delta s_e^2}{c_e^2} - 2\frac{s_e}{c_e}\frac{\Sigma^{\gamma Z}(0)}{m_Z^2} - \delta_{\rm VB} + \Delta\rho_f\right), \tag{1.200}$$

where $\Delta\rho_f$ denotes the finite part of the Zff vertex correction;
- the effective mixing angles of the Zff couplings:

$$s_f^2 = s_e^2(1 - \Delta\kappa_e + \Delta\kappa_f).$$

These relations predict the Z^0 boson couplings, m_W and ρ in terms of the data points $\alpha, G_\mu, m_Z, s_e^2$. By this procedure, the m_t^2 dependence of the self-energy corrections to the theoretical predictions is absorbed into the renormalized ρ parameter, leaving a term $\sim \log m_t/m_Z$ as an observable effect. For the Zbb vertex, an additional m_t^2 dependence is found in the non-universal vertex corrections $\Delta\rho_b$ and $\Delta\kappa_b$. This makes observables containing this vertex the most sensitive top indicators in the class of models with $\rho_{\rm tree} \neq 1$.

Given a specific model, one can calculate the value for ρ from

$$\rho = \frac{\pi\alpha}{\sqrt{2}G_\mu m_Z^2 s_e^2 c_e^2}\frac{1}{1-\Delta r} \tag{1.201}$$

in terms of the input data $\alpha, G_\mu, m_Z, s_e^2$, together with m_t and the parameters of the Higgs sector. Such a complete calculation has been performed for a

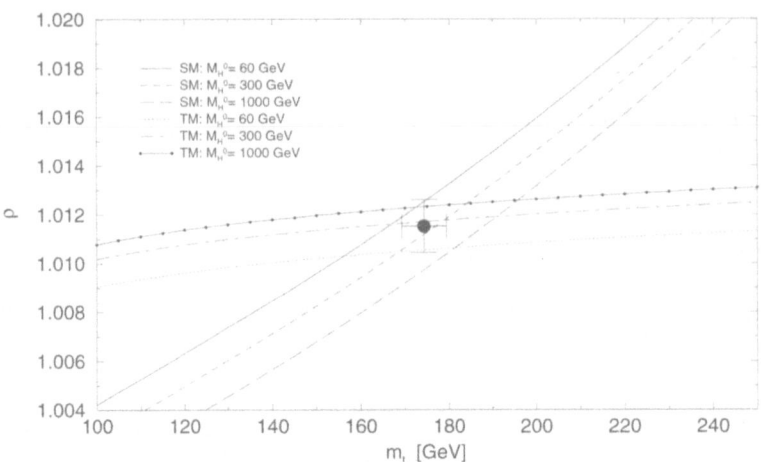

Fig. 1.8. The ρ parameter (1.197) in the Standard Model (SM) and in a model with an extra Higgs triplet (TM). The masses of the non-standard Higgs bosons for TM were set to 300 GeV. $\rho(TM)$ was calculated from (1.201) with the LEP/SLC data for s_e^2

model with an extra Higgs triplet, involving one extra neutral Higgs and a pair of charged Higgs bosons [122]. Figure 1.8 shows that the model predictions coincide in the area of the experimental data points with the values of m_t and ρ given by (1.197).

1.5.3 Extra Z Bosons

The existence of additional vector bosons is predicted by GUT models based on groups bigger than SU(5), such as E_6 and SO(10), by models with symmetry breaking in terms of a strongly interacting sector and by composite scenarios. Typical examples of extended gauge symmetries are the SU(2)×U(1)×U(1)$_{\chi,\psi,\eta}$ models following from E_6 unification, or left–right-symmetric models. In the following we consider only models with an extra U(1).

The mixing between the mathematical states Z_0 of the minimal-gauge group and Z_0' of an extra hypercharge forms the physical mass eigenstates Z, Z', where the lighter Z is identified with the resonance observed at LEP. The mass eigenstates are obtained by a rotation

$$Z = \cos\theta_{\mathrm{M}}\, Z_0 + \sin\theta_{\mathrm{M}}\, Z_0' ,$$
$$Z' = -\sin\theta_{\mathrm{M}}\, Z_0 + \cos\theta_{\mathrm{M}}\, Z_0' , \qquad (1.202)$$

with a mixing angle θ_{M} related to the mass eigenvalues by

$$\tan^2\theta_{\mathrm{M}} = \frac{m_{Z_0}^2 - m_Z^2}{m_{Z'}^2 - m_{Z_0}^2}, \quad m_{Z_0}^2 = m_Z^2 \cos^2\theta_{\mathrm{M}} + m_{Z'}^2 \sin^2\theta_{\mathrm{M}} . \qquad (1.203)$$

$m_{Z_0}^2$ denotes the nominal mass of Z_0. In constrained models with the Higgs fields in doublets and singlets only, the usual Standard Model relation

$$\sin^2\theta_W = 1 - \frac{m_W^2}{m_{Z_0}^2}$$

holds between the masses and the mixing angle in the Lagrangian

$$\mathcal{L}_{NC} = \frac{g_2}{2\cos\theta_W}\, J_{Z_0}^\mu\, Z_0^\mu + g'\, J_{Z_0'}^\mu\, Z_0'^{\,\mu} \tag{1.204}$$

with (1.24)

$$J_{Z_0}^\mu = 2(J_L^\mu - \sin^2\theta_W\, J_{em}^\mu)\,.$$

It is convenient to introduce the quantities

$$s_W^2 = 1 - \frac{m_W^2}{m_Z^2}, \quad c_W^2 = 1 - s_W^2\,, \tag{1.205}$$

where m_Z is the physical mass of the lower mass eigenstate. For small mixing angles θ_M we have the following relation:

$$\sin^2\theta_W = s_W^2 + c_W^2\Delta\rho_{Z'} \tag{1.206}$$

with

$$\Delta\rho_{Z'} = \sin^2\theta_M\left(\frac{m_{Z'}^2}{m_Z^2} - 1\right)\,. \tag{1.207}$$

The W mass is obtained from

$$m_W^2 = \frac{\pi\alpha}{\sqrt{2}G_\mu\sin^2\theta_W(1-\Delta r)}$$

after the substitution (1.206):

$$m_W^2 = \frac{m_Z^2}{2}\left(1 + \sqrt{1 - \frac{\pi\alpha}{\sqrt{2}G_\mu m_Z^2\rho_{Z'}(1-\Delta r)}}\right) \tag{1.208}$$

with $\rho_{Z'} = (1-\Delta\rho_{Z'})^{-1}$. Formally, $\rho_{Z'}$ appears as a non-standard tree-level ρ parameter. In all present practical applications the radiative correction Δr has been approximated by the Standard Model correction.

The mass mixing has two implications for the NC couplings of the Z boson.

- $\Delta\rho_{Z'}$ contributes to the overall normalization by a factor

$$\rho_{Z'}^{1/2} \simeq 1 + \frac{1}{2}\Delta\rho_{Z'}$$

and to the mixing angle by a shift

$$s_W^2 \to s_W^2 + c_W^2\Delta\rho_{Z'}\,.$$

Both effects are universal, parametrized by $m_{Z'}$ and the mixing angle θ_M in a model-independent way.

- A non-universal contribution is present as the second term in the vertex

$$(Zff) = \cos\theta_{\mathrm{M}}(Z_0 ff) + \sin\theta_{\mathrm{M}}(Z_0' ff)$$
$$\simeq (Z_0 ff) + \theta_{\mathrm{M}}(Z_0' ff).$$

This contribution depends on the classification of the fermions under the extra hypercharge and is strongly model-dependent.

Complete one-loop calculations are not available as yet. The present standard approach consists in the implementation of the Standard Model corrections to the Z_0 parts of the coupling constants in terms of the form factors ρ_f for the normalization and κ_f for the effective mixing angles:

$$s_W^2 \to s_f^2 = \kappa_f s_W^2.$$

In this approach the effective Zff vector and axial-vector couplings read

$$v_Z^f = \left[\sqrt{2} G_\mu m_Z^2 \rho_f (1 + \Delta\rho_{Z'})\right]^{1/2} \left[I_3^f - 2Q_f(\kappa_f s_W^2 + c_W^2 \Delta\rho_{Z'})\right]$$
$$+ \sin\theta_{\mathrm{M}} v_{Z_0'}^f,$$

$$a_Z^f = \left(\sqrt{2} G_\mu m_Z^2 \rho_f\right)^{1/2} I_3^f + \sin\theta_{\mathrm{M}} a_{Z_0'}^f. \tag{1.209}$$

The quantities $a_{Z_0'}^f, v_{Z_0'}^f$ denote the extra U(1) couplings between the fermion f and the Z_0'. The analyses of the electroweak precision data [123] constrain the mixing angle, typically to $|\theta_{\mathrm{M}}| < 0.01$.

1.5.4 The Minimal Supersymmetric Standard Model (MSSM)

Among the extensions of the Standard Model, the MSSM [124] is the theoretically favoured scenario, as the most predictive framework beyond the Standard Model. A definite prediction of the MSSM is the existence of a light Higgs boson with mass below ~ 135 GeV [125]. The detection of a light Higgs boson at LEP could be a significant hint of supersymmetry.

The structure of the MSSM as a renormalizable quantum field theory allows a complete calculation of the electroweak precision observables similar to that for the Standard Model, in terms of one Higgs mass (usually taken as the CP-odd "pseudoscalar" mass m_A) and $\tan\beta = v_2/v_1$, together with the set of SUSY soft-breaking parameters fixing the chargino/neutralino and scalar fermion sectors. It has been known for quite some time [120,126] that light non-standard Higgs bosons as well as light top squarks and charginos predict larger values for the ratio R_b [127,129]. Complete one-loop calculations are available for Δr [128] and for the Z^0 boson observables [129].

A possible mass splitting between the bottom and top squarks $\tilde{b}_L$ and $\tilde{t}_L$, the supersymmetric scalar partners of the left-handed components of the bottom and top quark, yields a contribution $\Delta\rho_{\tilde{b}\tilde{t}}$ to the ρ parameter, according to (1.189), of the same sign as the standard top term. As a universal

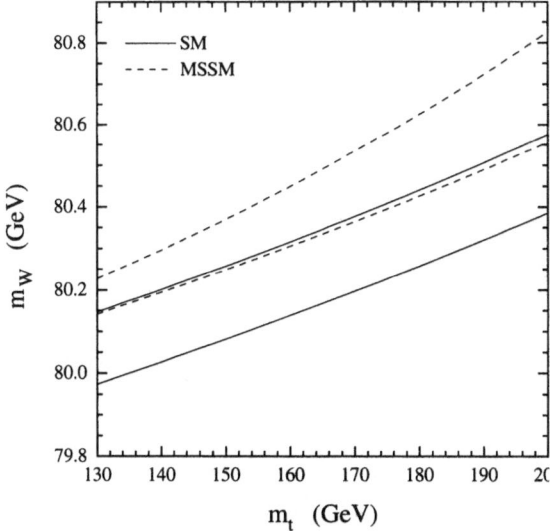

Fig. 1.9. The W mass range in the Standard Model (——) and in the Minimal Supersymmetric Standard Model (- - -). The bounds are from the non-observation of Higgs bosons and SUSY particles at LEP2

loop contribution, it enters the quantity Δr and the Z^0 boson couplings and is thus significantly constrained by the data on m_W and the leptonic widths. Recently the two-loop α_s corrections have been computed [131], which can amount to 30% of the one-loop $\Delta\rho_{\tilde{b}\tilde{t}}$.

Figure 1.9 displays the range of predictions for m_W in the minimal model and in the MSSM. It is assumed here that no direct discovery has been made at LEP2. As can be seen, precise determinations of m_W and m_t may become decisive for a separation between the models.

As the Standard Model, the MSSM yields a good description of the precision data. A global fit [130] to all electroweak precision data, including the top mass measurement, shows that the χ^2 of the fit is slightly better than in the Standard Model; but, owing to the larger numbers of parameters, the probability is about the same as for the Standard Model (Figure 1.10).

The virtual presence of SUSY particles in the precision observables can be also exploited in another way, of constraining the allowed range of the MSSM parameters. Since the quality of the Standard Model description can be achieved only for those parameter sets where the Standard Model with a light Higgs boson is approximated, deviations from this scenario result in a rapid decrease of the fit quality. An analysis of the precision data in this spirit can be found in [132].

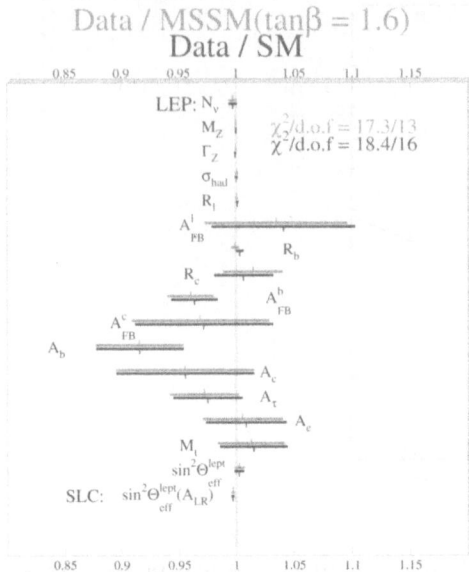

Fig. 1.10. Best fits in the Standard Model and in the MSSM, normalized to the data. The error bars are those from the data. (Updated from [130])

2. Precision Measurements

The properties of the Z^0 and $W^\pm$ vector bosons, their masses and couplings, form the key to today's understanding of elementary particles and their interactions. Precise measurements of these properties were the main motivation for the construction of the electron–positron colliders LEP and SLC.

Most of this chapter is devoted to the description of measurements done at the e^+e^- collider LEP at energies close to the Z^0 pole. During this so-called LEP1 phase (1989–1995) the four LEP experiments collected a huge sample of Z^0 decays; each experiment recorded about 4×10^6 hadronic events and 4×10^5 charged lepton pairs. This large data volume allows measurements with unprecedented precision, but it also imposes high demands on the experimental analyses. In order to exploit the statistical precision of the data, analysis techniques must be developed which reduce systematic uncertainties to the level of a few per mille. Only now, several years after the end of LEP1, have most of the measurements reached the final stage. Many of the results presented are still preliminary, corresponding to the status in summer 1999.

In addition, the electroweak measurements at the SLC e^+e^- collider and at LEP2 are summarized and the results from the TEVATRON $p\overline{p}$ collider and neutrino–nucleon scattering are briefly described.

The structure of this chapter is as follows. In Sect. 2.1 the LEP collider and experiments are introduced. Section 2.2 describes the lineshape measurements, i.e. the energy dependence of cross-sections and asymmetries. These form the core of the electroweak precision results, and the ingredients needed, LEP centre-of-mass energy, luminosity, event selections and lineshape fit, are discussed in some detail. Section 2.3 describes the polarization measurements made at LEP for tau final states and at SLC with polarized electron beams. An overview of the analysis methods and results for the determination of heavy-quark partial Z^0 decay widths and asymmetries is given in Sects. 2.4 and 2.5. In Sect. 2.6 the status of the measurements at LEP2 on the W boson mass and the search for the Higgs boson is briefly described. Finally, Sect. 2.7 summarizes electroweak results from other experiments, the mass of the top quark and the W boson from the TEVATRON and the determination of the electroweak mixing angle in neutrino–nucleon scattering.

2.1 LEP Collider and Detectors

The Large Electron Positron Collider LEP at CERN is the world's largest particle accelerator, with a circumference of almost 27 km. It is located at the Swiss–French border between Lake Geneva and the Jura Mountains about 100 m below the surface.

The LEP ring consists of eight arcs interspersed with eight straight sections. Electrons and positrons circulate along the ring in a single beam pipe, packed in four or eight bunches of typically 10^{11} particles. About 400 dipole magnets bend the beams in each of the arcs, additional quadrupole and sextupole magnets provide the focusing. Radio-frequency cavities, located in the straight sections, accelerate the particles and compensate the energy loss due to synchrotron radiation. At LEP1 energies, 128 copper cavities were sufficient to drive the beams. For LEP2 the power needed to compensate the synchrotron radiation is higher by a factor of 16. Therefore, the system was upgraded with high-gradient superconducting cavities, in addition to or replacing the old copper cavities.

The beams collide in four of the straight sections, where the LEP experiments are located. At the other beam crossing points collisions are avoided either by electrostatic separators or by specific beam optics which cause oscillations around the nominal orbit with opposite phase for electrons and positrons. A schematic map of LEP and its detectors is shown in Fig. 2.1.

LEP started operation in summer 1989. In the first phase (LEP1), from 1989 to 1995, it was operated with centre-of-mass energies close to the Z^0 resonance at 91 GeV. The second phase (LEP2) began in 1996 with energies above the W^+W^- pair production threshold at 161 GeV and will eventually extend until 2000 and reach a centre-of-mass energy above 200 GeV. The performance of the accelerator was steadily improved throughout the years and the design luminosity [133] of 1.6×10^{31} cm^{-2}s^{-1} was routinely achieved or even surpassed in the later years of LEP1.

During LEP1 the total luminosity recorded by each experiment was about 170 pb^{-1}; roughly 75 % of it was taken at the maximum of the Z^0 resonance. Altogether, the four experiments collected more than 16×10^6 Z^0 decays. At LEP2 data corresponding to an integrated luminosity of about 270 pb^{-1} above 160 GeV has been recorded by each experiment to date. A more detailed overview of the LEP operation is given in Table 2.1.

2.1.1 The LEP Experiments

All four LEP detectors – ALEPH, DELPHI, L3 and OPAL – are general-purpose detectors, i.e. they are designed to record all types of interactions which occur in e^+e^- collisions at LEP1 and LEP2 energies. The general layout of each detector follows the generic design of modern general-purpose detectors at high-energy colliders. In a cylindrical geometry around the beam pipe, there are the following.

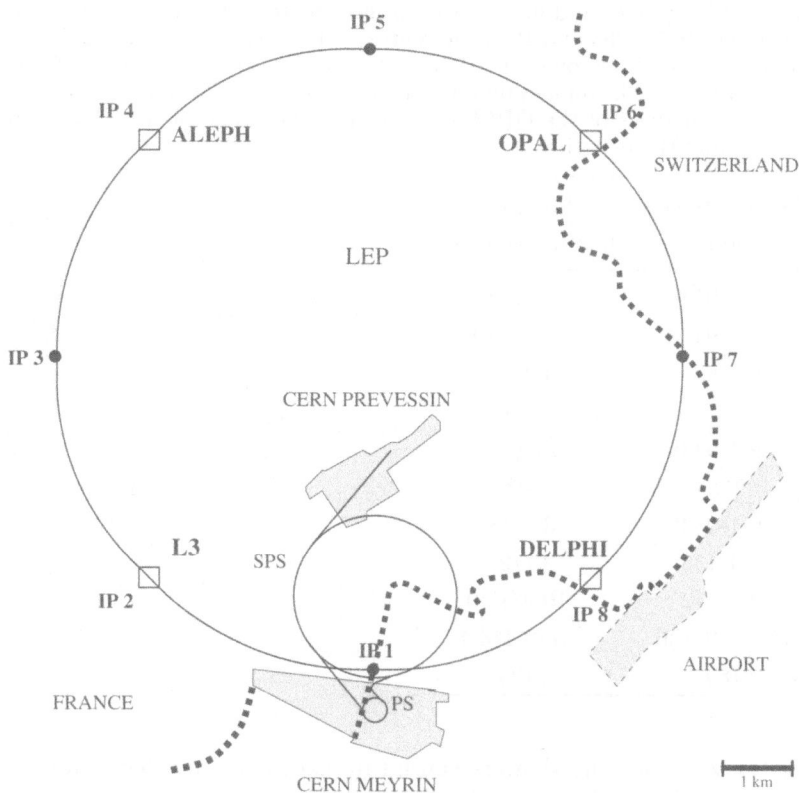

Fig. 2.1. Map of the LEP ring and the detectors

- Tracking chambers to reconstruct the trajectories of charged particles. A magnetic field oriented along the beam axis allows the measurement of the momentum of the particles from the curvature of the tracks. Particles created in the primary interaction or a secondary decay can be distinguished by extrapolating the tracks back to the origin. High-resolution silicon microvertex detectors are crucial to achieving a precise secondary-vertex reconstruction. In addition, by measuring the specific energy loss in the tracking volume, particles of different mass can be identified.
  All experiments use several layers of different tracking detectors. The innermost layer is a small, high-precision silicon microvertex detector, located as close as possible to the beam pipe, followed by one or more large-volume gas chambers.
- Calorimeters which absorb photons, electrons and hadrons and measure their energy. All LEP experiments use two layers: first an electromagnetic calorimeter to measure precisely the fast-evolving electromagnetic shower induced by photons and electrons, and then a much larger hadronic

Table 2.1. LEP energies and luminosities from 1989 to 1998. For the "Z^0 scan" years 1989, 1990, 1991, 1993 and 1995 the centre-of-mass energy was varied around the maximum of the Z^0 resonance; the energy range is given, and the off-peak luminosity in brackets. The quoted luminosities correspond to the data sample used for the lineshape analysis of the OPAL experiment. They vary slightly depending on the analysis and the experiment

	Centre-of-mass energy	Integrated luminosity (pb^{-1})
1989	88.3–95.0	1.3 (0.9)
1990	88.2–94.2	7.0 (3.5)
1991	88.4–93.7	13.0 (5.5)
1992	91.3	25
1993	89.4–93.0	32 (18)
1994	91.2	53
1995	89.4–93.0	32 (18)
	130/136	2.7/2.6
1996	161/172	10.0/10.0
1997	183/130/136	57/2.4/3.4
1998	189	195

calorimeter to record the showers caused by hadronic particles in nuclear interactions, which evolve more slowly

• Muon chambers as the outermost layer. In general only muons escape the calorimeters; they are identified by simple tracking chambers.

The tracking chambers are wrapped around most of the solid angle; typically they are sensitive within polar angles to the beam pipe of $20° < \theta < 160°$. The cylindrical part of the calorimeters and muon chambers extends over the range $40° < \theta < 140°$. It is complemented by two "end cap" sections which consist of a similar sequence of subdetectors. In addition, forward detectors, placed at small polar angles on both sides, are installed to measure the luminosity. As an example, the OPAL detector is shown in Fig. 2.2.

In detail, however, the LEP detectors differ substantially in their construction, both in the technologies used for the various components and in the emphasis put on specific properties. A summary of the dimensions, components and performance is given in Table 2.2. The main distinguishing features of the four detectors are as follows.

• ALEPH (A detector for LEP PHysics [134]) uses a large cylindrical time projection chamber (TPC) as a central tracking detector. It has a good resolution both perpendicular to the beam direction ($r\Phi$) and parallel to it (z). Together with the vertex detectors and the strongest superconducting

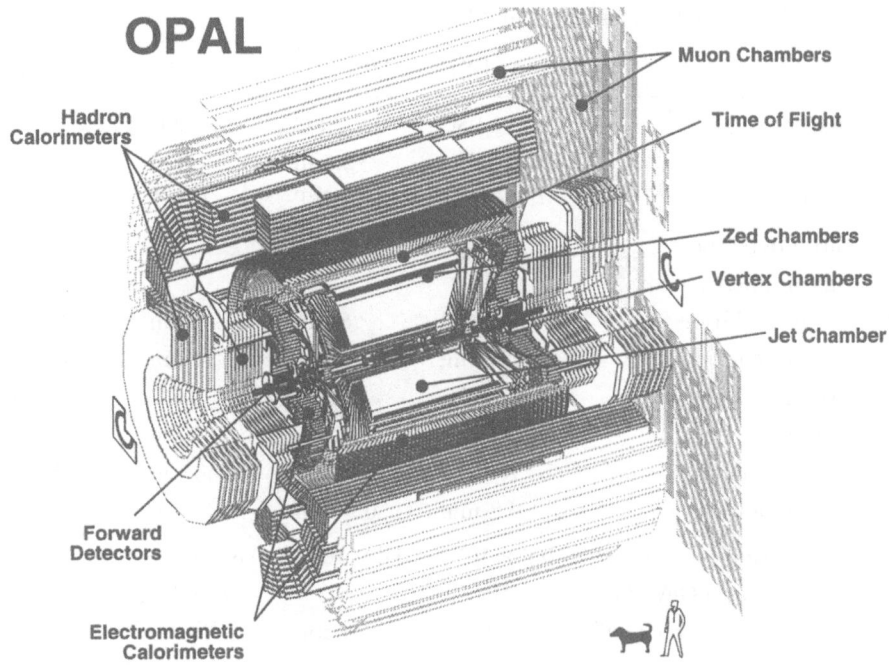

Fig. 2.2. The OPAL detector at LEP

solenoid of all LEP detectors it obtains an excellent momentum resolution. Also inside the solenoid, a sandwich electromagnetic calorimeter with alternating layers of lead and proportional tubes is situated. The main emphasis of the design is not energy resolution but a fine granularity, which is achieved by small transverse readout pads and an additional longitudinal segmentation, which is advantageous for shower separation and particle identification.

- DELPHI (DEtector with Lepton, Photon and Hadron Identification [135]) employs the most ambitious design in terms of novel detector technologies. The particular feature is ring-imaging Cerenkov detectors, which allow particle identification up to much higher momenta than the conventional energy loss measurements in tracking chambers. Similarly to ALEPH, DELPHI uses a superconducting solenoid outside a fine-granularity electromagnetic calorimeter and a large TPC for the central tracking.
- L3 (3rd Lep detector [136]) differs most from the other detectors in the dimensions of the individual subdetectors. It has a huge magnet surrounding all the other components. The precise muon chambers form an additional spectrometer, which yields an excellent muon momentum resolution. This layout enforces small dimensions for the central tracking chambers and the calorimeters. On the one hand this compromises the momentum resolution for electrons and hadrons and limits the particle identification in

Table 2.2. Parameters of the four LEP detectors

	ALEPH	DELPHI	L3	OPAL
Magnet	SC	SC	NC	NC
radius	2.6 m	2.6 m	6 m	2.2 m
field	1.5 T	1.23 T	0.5 T	0.435 T
Central detector				
silicon microvertex				
$r\Phi$ resolution	12 μm	8 μm	7 μm	5 μm
z resolution	10 μm	9 μm	14 μm	13 μm
tracking chamber	TPC	TPC	TEC	Jet
radius	1.8 m	1.2 m	0.45 m	1.8 m
$r\Phi$ resolution	180 μm	250 μm	50 μm	135 μm
z resolution	0.8 mm	0.9 mm	–	45 mm
momentum resolution	0.6×10^{-3}	0.6×10^{-3}	0.6×10^{-3}	1.3×10^{-3}
$\sigma(p_{\mathrm{T}})/p_{\mathrm{T}}^2$ (GeV^{-1})			(μ only)	
impact parameter ($r\Phi$)	23 μm	20 μm	30 μm	16 μm
Calorimeter				
electromagnetic	Lead / P.T.	HPC	BGO	Lead glass
granularity ($\theta \times \phi$)	$0.9° \times 0.9°$	$1.0° \times 1.0°$	$2.3° \times 2.3°$	$2.3° \times 2.3°$
energy resolution	$0.18/\sqrt{E}$	$0.32/\sqrt{E}$	$0.02/\sqrt{E}$	$0.06/\sqrt{E}$
σ_E/E	$\oplus 0.01$	$\oplus 0.04$	$\oplus 0.01$	$\oplus 0.01$
hadron				
energy resolution	$0.85/\sqrt{E}$	$1.12/\sqrt{E}$	$0.7/\sqrt{E}$	$1.2/\sqrt{E}$

the central detector, a time expansion chamber (TEC). On the other hand it makes the use of expensive bismuth germanium oxide (BGO) crystals for the electromagnetic calorimeter possible, which provides an excellent energy resolution.

• OPAL (Omni Purpose Apparatus for Lep [137]) is the most conservative construction of the four detectors. Deliberately, its design was based on existing technologies which were succesfully used before in other experiments. The main part of the central detector is a large jet chamber equipped with nearly 4000 precisely mounted signal wires, which gives both an excellent $r\Phi$ resolution and a very good measurement of the specific energy loss. The electromagnetic calorimeter consists of lead glass blocks, which have a high energy resolution but a relatively coarse granularity compared to ALEPH and DELPHI. A handicap is the conventional (normal-conducting) magnet. It provides a relatively low magnetic field, which compromises the momentum resolution. Furthermore, it is situated in front of the electromagnetic calorimeter, which causes pre-showering in the magnet coil and dilutes the energy measument.

The clean operating conditions at the LEP collider make it possible to impose only very loose requirements for the triggering and recording of events. Typically, genuine Z^0 events were produced at a rate of about 1 Hz at LEP1. The trigger systems combine the information from many subdetectors which gives a good redundancy and results in trigger efficiencies close to 100 % for events originating from Z^0 decays[1].

The Monte Carlo simulation plays a crucial role in the analysis of the recorded data. This is usually done in two steps, the event generation and the simulation of the detector response. For the simulation of the physics of the reaction numerous event generators are available for the different final states. In general these are common to all LEP experiments; only the specific tuning of model parameters to the observed properties is done individually by each experiment. The detector simulation requires first an accurate geometrical mapping of all detector components and materials. The interaction of traversing particles and the response of the subdetectors is then simulated in detail with the GEANT [138] package. The simulated events are processed by the same reconstruction and analysis chain as is used for the real data. In most analyses Monte Carlo events are indispensable for optimizing the selection criteria, determining correction factors and studying systematic effects.

2.2 Z^0 Lineshape and Asymmetry

In the Standard Model Z^0 bosons decay into fermion–antifermion pairs. There are eleven possible decay channels: five quark flavours (the top quark is too heavy), three charged leptons and three neutrino species, the latter escape the detector undetected. In the "standard" lineshape analysis no attempt is made to distinguish the different quark flavours (all $Z^0 \rightarrow q\bar{q}$ decays are classified as "hadronic" events), while the charged leptons (electron, muon and tau pairs) are selected separately.

Experimentally, three separate measurements are needed to determine the cross-section for a final-state fermion pair $f\bar{f}$, which is given by

$$\sigma_f(E_1) = \frac{N_f^{\text{cand}}(E_i) - N_f^{\text{bg}}(E_i)}{\varepsilon_{\text{acc}}^f(E_i) \int \mathcal{L}\, dt} \ . \tag{2.1}$$

- The event selection counts the number of events, N_f^{cand}, at a specific centre-of-mass energy point and must determine precisely the corrections which account for the experimental detection efficiency, $\varepsilon_{\text{acc}}^f$, and the expected background contamination, N_f^{bg}.

---

[1] Except $Z^0 \rightarrow \nu\bar{\nu}$ decays, of course.

- The integrated luminosity $\int \mathcal{L}\,dt$ must be determined at this energy point.
- The centre-of-mass energy of LEP, E_i, must be measured precisely.

The huge data statistics recorded at LEP1 set the scale for the precision required for these measurements. About four million hadronic decays and 150 000 events in each of the lepton categories correspond to a statistical precision of 0.05 % and 0.25 %, respectively. Ideally, this should be matched by the selection and luminosity uncertainties.

A substantial fraction of the data was taken at centre-of-mass energies below or above the Z^0 resonance. In order to fully exploit the achievable statistical precision for m_Z and Γ_Z it is required to determine the LEP centre-of-mass energy at the level of 1 MeV.

In the following subsections the LEP energy calibration, the luminosity measurement and the event selections are discussed.

2.2.1 LEP Energy Determination

The precise determination of the mass and width of the Z^0 boson is based on a scan of the centre-of-mass energy around the pole of the Z^0 resonance. In the first two years of LEP, 1990 and 1991, scans were performed at seven energy points between 88 and 94 GeV: three points below and above the peak and one point at the maximum of the resonance. In 1993 and 1995 short scans were done with only three energy points, at the peak and about 1.8 GeV below and above. In 1992 and 1994 LEP was operated at the peak energy.

Naturally, the precision of the Z^0 mass and width measurements depends on the accuracy of the centre-of-mass energy. The mass is directly related to the knowledge of the absolute energy scale, which is strongly correlated for the different scan points. In contrast, the width is mainly affected by the relative uncertainty between energy points, the independent part of the uncertainty in the beam energy.

For the 1993 and 1995 scans, which completely dominate the overall measurement, the errors on the mass and width caused by the centre-of-mass energy uncertainties are approximately

$$\Delta m_Z \approx 0.5\Delta(E_{+2} + E_{-2})\,, \tag{2.2}$$

$$\Delta \Gamma_Z \approx 0.71\Delta(E_{+2} - E_{-2})\,. \tag{2.3}$$

E_{-2} and E_{+2} are the centre-of-mass energies at the two off-peak points, about 2 GeV below and above the Z^0 peak.

The pure statistical accuracy when combining the data from all four LEP experiments is about 1.3 MeV for m_Z and 2.0 MeV for Γ_Z.

A further, more subtle effect which must be taken into account when the cross-sections are calculated is the beam energy spread, i.e. the scatter of the energies around the nominal beam energy, of typically 50 MeV. Therefore the measured cross-section does not correspond to a sharp centre-of-mass energy but effectively to a weighted average over the range given by the energy spread.

Calibration

The cornerstone of the LEP energy calibration is the resonant spin depolarization, a method which exploits the extremely precise knowledge of the electron magnetic-moment anomaly, $g - 2$, to measure the average energy of the beam in the LEP ring with an intrinsic accuracy well below 1 MeV.

The resonant depolarization is a rather delicate measurement and it could not be achieved under standard physics data-taking conditions. Instead it was performed routinely at the end of the LEP fills with separated beams and special accelerator settings.

Therefore one needs to extrapolate from the resonant depolarization measurements to the average beam energy under physics conditions. Besides several anticipated corrections for this extrapolation, a number of unexpected effects were found in the analysis, the most spectacular being the energy variation induced by tidal effects and the influence of parasitic currents caused by electric trains in the Geneva area.

Furthermore, the centre-of-mass energy at the interaction points (IPs) is not necessarily twice the average beam energy. Depending on the positioning and the configuration of the accelerating cavities, IP-specific changes of up to 20 MeV can occur. In addition, energy dispersion can affect the centre-of-mass energy.

In the following the LEP energy calibration is briefly described, with an overview of the instruments and methods used, and then the various effects influencing the centre-of-mass energy and the results are discussed. Many more details can be found in the reports and publications of the LEP energy working group [139].

Instrumentation

The absolute scale of the LEP beam energy is determined very accurately by resonant depolarization. But in order to extrapolate to the effective centre-of-mass energy at each IP, information from many additional sources is needed: magnet currents, temperatures and fields, beam orbit measurements and cavity set-up. A short overview of these instruments is given below.

Resonant Depolarization. Transverse spin polarization builds up in a storage ring because of the Sokolov–Ternov effect [140]. It is caused by the emission of synchrotron radiation, which favours the alignment of the electron spin opposite to the bending field. Under ideal conditions in a homogeneous dipole field a transverse polarization of 95 % can be expected. In practice, the polarization build-up is strongly affected by field distortions and the experimental solenoids. Only after careful adjustments and compensations can transverse polarization be observed, and a level of 10–20 % is regularly reached at LEP under these special conditions.

Transverse polarization is an ideal tool for the beam energy measurement, since the number of spin precessions per turn is directly related to the beam energy:

$$\nu_{\mathrm{s}} = \frac{1}{2}(g_e - 2)\gamma\,, \tag{2.4}$$

where $g_e - 2$ is the anomalous magnetic moment of the electron and $\gamma = \frac{E_{\mathrm{beam}}}{m_e}$ the relativistic Lorentz factor. Both $g_e - 2$ and m_e are known with a precision better than 1 ppm. The beam energy can be written

$$E_{\mathrm{beam}} = \frac{\nu_{\mathrm{s}} m_e}{(g_e - 2)/2} = \nu_{\mathrm{s}} \times 440.6486(1) \text{ MeV}\,. \tag{2.5}$$

In order to measure the spin tune ν_{s} an oscillating magnetic field in the radial direction is applied, which depolarizes the beam when its frequency (f_{dep}) matches the product of the fractional spin tune and the revolution frequency (f_{rev}) of the beam:

$$f_{\mathrm{dep}} = [\nu_{\mathrm{s}} - \text{int}(\nu_{\mathrm{s}})]\, f_{\mathrm{rev}}\,. \tag{2.6}$$

The polarization itself is monitored with a laser polarimeter. This method is based on the spin dependence of Compton scattering of circularly polarized laser light off the electron beam. The angular distribution of backscattered photons depends on the relative orientation of the photon polarization and the electron spin. Alternating left- and right-handed circular polarization provides a sensitive measure of the beam polarization. A typical polarization experiment is shown in Fig. 2.3, which demonstrates the high intrinsic resolution of about 200 keV.

But in reality several complications arise. Depolarization occurs not only at the nominal spin tune $\delta_{\mathrm{s}} = \nu_{\mathrm{s}} - \text{int}(\nu_{\mathrm{s}})$ but also at the mirror frequency $1 - \delta_{\mathrm{s}}$ and at sidebands $\delta_{\mathrm{s}} \pm Q_{\mathrm{s}}$.[2] Then the polarization measurement alone is ambiguous, as it is sensitive only to the fractional spin tune. A beam energy different by 440.6 MeV gives depolarization at the same frequency. However, many other instruments exist that can resolve the beam energy at this level. Furthermore, to reach a sufficient level of polarization, half-integer spin tunes are preferable. This dictates the settings for the LEP centre-of-mass energy in steps of 881 MeV. It is a lucky coincidence that the Z^0 peak at 91.2 GeV corresponds to a half-integer, with $\nu_{\mathrm{s}} \approx 103.5$.

Magnetic-Field Monitoring. In order to track the energy changes from a resonant depolarization measurement to any point in time during physics data-taking one uses several devices to monitor the magnetic dipole field:

---

[2] Q_{s} is the synchrotron tune, the number of longitudinal or energy oscillations per turn.

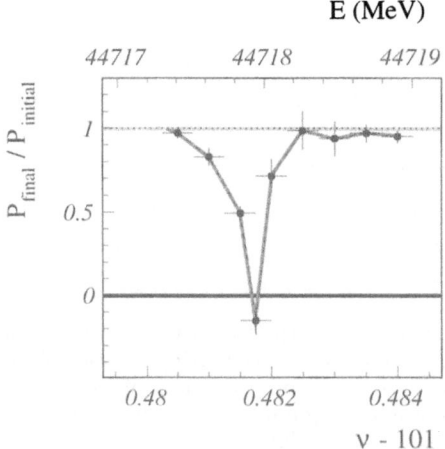

Fig. 2.3. Example of a resonant-polarization measurement. The frequency of the depolarizing field, $f_{\rm dep}$, is slowly varied and at a certain frequency the polarization disappears. This frequency can be directly converted into the beam energy

1. measurement of the dipole current;
2. NMR (nuclear magnetic resonance) probes in a reference magnet, which is powered in series with the LEP magnets;
3. NMR probes in two LEP magnets (only installed in 1995).

In principle, with an accurate knowledge of the magnetic field, the beam energy can be determined by integrating the magnetic field along the beam orbit:

$$E_{\rm beam} \propto \oint B dl \,. \tag{2.7}$$

However, all these devices give only an approximate picture of the magnetic field seen by the beam. Obviously, the measurement of the current cannot track hysteresis effects and variations caused by temperature changes. Similarly, the reference magnet cannot reproduce all changes, as it is a different magnet construction in a different environment. Much better is the information from the NMR probes in the LEP tunnel, but they still only give a spot test.

Complementary tools are needed to allow the tracking of the centre-of-mass energy with sufficiently high precision.

Beam Orbit Monitors. Most of the LEP quadrupoles are equipped with beam orbit monitors (BOMs), consisting of four capacitive pick-up plates which measure the beam orbit relative to the quadrupole centre with a precision of a few microns. With this information one can control deformations of the LEP ring caused, for example, by tidal or hydrogeological effects. It also serves as an additional cross-check of the settings of the accelerating cavities.

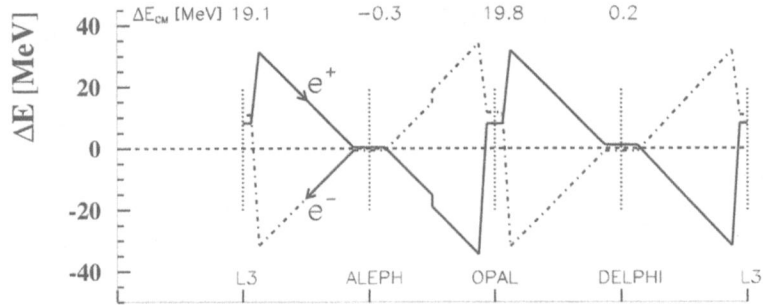

Fig. 2.4. Deviation of the average beam energy along the LEP ring. Owing mostly to synchrotron radiation, the energy becomes reduced. Accelerating cavities placed around L3 and OPAL compensate the loss. Owing to misalignment of the cavities the energy gain is asymmetric for incoming and outgoing beams, which causes an net increase in the centre-of-mass energy at L3 and OPAL

RF Cavities. At LEP1 energies the electrons and positrons lose about 125 MeV per turn owing to synchrotron radiation. This is compensated by radio-frequency (RF) copper cavities which are located in the straight sections of the LEP ring close to the OPAL and L3 experiments. Figure 2.4 shows the variation of the average beam energy due to synchrotron radiation and the acceleration along the LEP ring. Ideally the cavities next to L3 and OPAL should provide just the right amount of energy to the beams to ensure that the centre-of-mass energy at the four interaction points is the same. However, in 1992 it was found that the cavities are systematically misaligned by about 2.5 cm with respect to the interaction points.[3]

As a result the incoming beams in OPAL and L3 acquire about 10 MeV more energy and the outgoing beams about 10 MeV less. Therefore the effective centre-of-mass energy is about 20 MeV larger in OPAL and L3. The exact size of the shift is rather sensitive to the details of the cavity power distribution.

Effects on the LEP Centre-of-Mass Energy

Deformations of the LEP Ring. The electrons and positrons in LEP are ultrarelativistic, hence their speed is constant to a very high degree. Since the RF frequency is fixed the length of the beam orbit is constant, i.e. independent of the beam energy.[4] Therefore any deformation of the LEP tunnel changes the position of the beam relative to the magnets. In the "central

---

[3] For technical reasons the copper cavities are operated at two frequencies $f_1 =$ 352.209 MHz and $f_2 = $ 352.299 MHz. For symmetric operation the cavities on the two sides of the interaction point should be placed at multiples of $\lambda_{RF} =$ $2\,c/(f_1 + f_2)$ from the IP. Instead the alignment was done for c/f_1.

[4] The speed of the electrons changes by less than 10^{-11} when raising the centre-of-mass energy from 88 to 92 GeV.

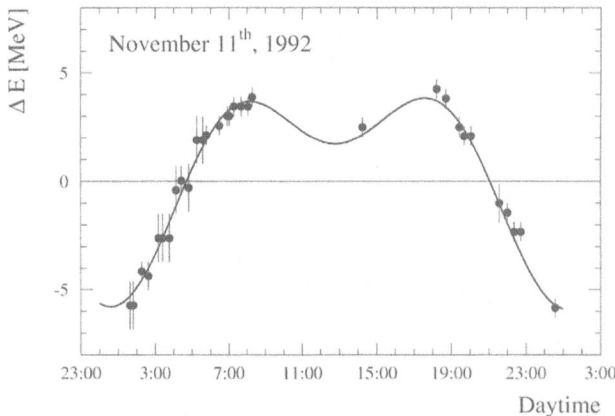

Fig. 2.5. Variation of the beam energy due to tidal effects. The results of polarization measurements are indicated as *points with error bars*. The *line* shows the theoretical prediction

orbit" the particles pass on average through the centre of all quadrupole and sextupole magnets, where the magnetic field is zero. But as soon as the relative position of the particle orbit changes, a residual magnetic field contributes to the dipole bending field, which makes the beam energy rather sensitive to deformations of the LEP ring. The strong focusing optics of LEP cause a large amplification of radial changes

$$\frac{\Delta E}{E} = \frac{1}{\alpha_c} \frac{\Delta R}{R} \,, \tag{2.8}$$

where $\alpha_c = 1.86 \times 10^{-4}$ is the momentum compaction factor.

Tidal forces exerted by moon and sun give rise not only to the tides in the oceans but also to tides in the earth's crust with an amplitude of about 30 cm. This oscillation induces a deformation of the ground affecting the LEP ring at the 10^{-8} level, which corresponds to a 7 MeV variation of the beam energy.

This effect was originally not anticipated but its impact was soon realized after the first year of precision resonant-depolarization measurements in 1991 [141]. Now it is a well-established correction, which is both accurately predicted by theoretical models and nicely reproduced in dedicated measurements (Fig. 2.5). Small uncertainties in possible phase shifts and the deformation amplitude contribute about 0.2 MeV to the centre-of-mass energy error.

Moreover, hydrogeological effects cause long-term drifts of the beam orbit. A clear correlation could be observed between the water level in the Lake of Geneva and the BOM measurements. Heavy rainfall in the neighbouring Jura mountains acts similarly. Together these effects can change the beam energy by up to 20 MeV, but the drifts are very slow (≈ 1 month) and well tracked

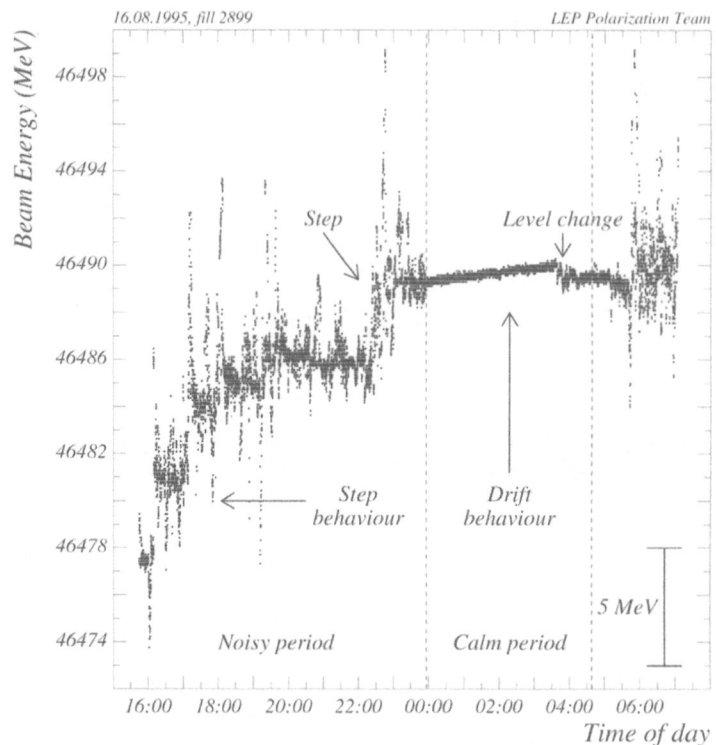

Fig. 2.6. Rise of the magnetic dipole field (measured with NMR and converted into beam energy units) during a LEP fill

by the BOM system and the resonant depolarization; therefore the error is negligible.

Variations of the Magnetic Dipole Field. The resonant depolarization calibration was mostly performed at the end of the LEP fills. Therefore one needs to know precisely any change in the average dipole field in order to extrapolate the beam energy to the preceding physics data-taking. This turned out to be a rather complicated matter and became the most important issue of the LEP energy calibration for the precision scan years 1993 and 1995.

The two NMR probes installed in two dipoles in the LEP tunnel in 1995 showed a systematic rise of the field during the fills (Fig. 2.6). This rise, corresponding to typically 8 MeV in the beam energy during a fill, was not observed in the reference magnet. However, in six fills resonant depolarization was performed both at the beginning and at the end of the fill and a similar rise of the energy was seen.

One source of the rise is parasitic currents through the LEP beam pipe. These had a very particular time structure – only from 5:00 a.m. to 12:00 p.m. – and were flowing from the southern part of the ring in both directions

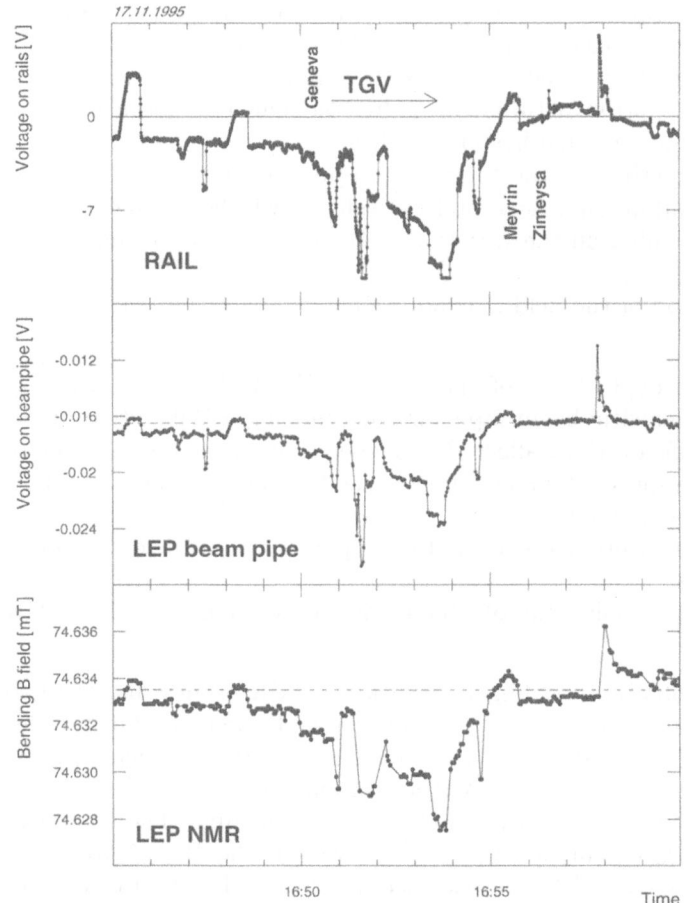

Fig. 2.7. Correlation between voltage measured on rails caused by a departing TGV express, voltage induced on the LEP beam pipe and magnetic dipole field measured with NMR probes

to the north-eastern part. These patterns finally helped to identify the source: electric trains arriving at and leaving Geneva south of the LEP ring cause a leakage current to the ground which finds its way to the LEP ring and returns via a small river north-east of LEP to the power station. A dedicated experiment nicely demonstrated the correlation between a departing TGV express, the beam pipe current and the dipole field as measured by NMR (Fig. 2.7).

The currents flowing in the ring affect the magnetization of the iron in the dipole magnets; a current spike gives a net change in the magnetization due to hysteresis effects, which also explains the apparent saturation of the field rise seen in Fig. 2.6.

Moreover, temperature variations also modify the dipole field. Firstly, thermal expansion and contraction gives rise to geometric changes. But more important are thermally induced stresses at the contact zone between the iron and the concrete support structure, which alter the magnetic permeability of the iron and introduce thermal hysteresis effects.

In dedicated experiments the temperature dependence of the magnetic field was studied and an empirical model developed which parametrizes the magnetic field depending on the variation of the temperature during operation.

For the correction of the field rise, additional complications were caused by the following.

- The temperature dependence of the dipole field interferes with the rise induced by the parasitic beam pipe currents, making it difficult to disentangle the two effects. Both affect the magnetization hysteresis; therefore it depends on the specific temperature and daytime history of each fill how much each effect contributes.
- No direct NMR measurements of the LEP dipole magnetic field were done in 1993 and 1994.
- In 1995 only two dipoles out of about 3400 were equipped with NMR probes.

It was verified in 1996 by installing 14 additional NMR probes that the two 1995 probes reproduced reasonably well the average behaviour. But in order to correct the field rise in 1993 and 1994, models are needed which describe the field rise independently of the NMR measurements.

To control the systematics two models were compared. The first uses the detailed temperature model discussed above and the 1995 NMR measurements to parametrize the parasitic-current effect as function of time of day, assuming the independence of the effects. The second uses a detailed parasitic-current recording over one week of LEP running and a dedicated measurement of the combined temperature and parasitic-current effects. The two models give reasonable agreement and also reproduce the beam energy rise as measured by the six fills with resonant depolarization at the beginning and end of the fill.

Overall, the dipole field rise results in an uncertainty in the centre-of-mass energy of 2.7 MeV for 1993 and 1.2 MeV for 1995. The latter is much smaller as more information from the resonant depolarization and NMR measurements was available, which gives better constraints and more powerful consistency checks.

RF Correction. The systematic misalignment of the RF copper cavities which are situated close to the OPAL and L3 interaction regions causes a substantial shift of up to 20 MeV in the collision energy for these experiments.

The specific size of the shift depends on the cavity configuration, i.e. the distribution of the power in the cavities and their phases. In particular, for the 1995 running when new (correctly aligned) superconducting cavities were commissioned, which partially replaced the copper system, a careful analysis of the RF configuration was crucial. '

A simulation model was developed which calculates the stable phase angle, Φ_{RF}, of the bunches on the RF wave. In a first approximation this can be obtained by setting the energy loss by synchrotron radiation per turn equal to the gain in the RF cavities:

$$\sum V_{RF} \sin \Phi_{RF} = E_{sync} \approx 125 \text{ MeV} . \tag{2.9}$$

Additional corrections due to the beam currents, unused cavities, RF phase shifts and wigglers need to be taken into account. The RF model gives not only Φ_{RF} but also the synchrotron tune Q_s, the longitudinal position of the interaction point and the difference in radius between the electron and positron beam orbits. These quantities can be compared with the direct measurement of Q_s, the determination of the event vertex in the experiments and the monitoring of the orbits with the BOM system. Together this information gives powerful constraints on the RF cavity configuration, sufficient to control systematic uncertainties at the level of 0.5 MeV in 1993 and 0.7 MeV in 1995.

Dispersion Effects. In 1995 LEP was operated in "bunch train mode". In this mode four equally spaced "trains" (≈ 6.7 km apart) containing 2–4 bunches each (74 m apart) circulate in the LEP collider.[5] To avoid collisions in the parasitic interaction points electrostatic separators were used to separate the electron and positron beam vertically. This separation introduces a vertical dispersion of the beams, i.e. the beam energy depends on the position of the particle in the beam. This dispersion has opposite signs for electrons and positrons. Hence any vertical offset in the collision causes a change of the effective centre-of-mass energy with a typical size of 1.5 MeV per 1 μm displacement. The effect was anticipated beforehand and a careful online monitoring and adjustment of the optics kept the average vertical offset mostly below 0.5 μm, which led to small corrections and only a small uncertainty of 0.5 MeV.

Beam Energy Spread. Electrons and positrons circulating in a storage ring perform oscillations around their nominal RF phase angle, the synchrotron oscillations. These are driven by the synchrotron radiation: statistical fluctuations in the quantized emission of synchrotron photons excite the oscillation while the energy dependence of the synchrotron loss causes the damping. Therefore the energy in the beam is smeared around the nominal energy, with a Gaussian shape due to the statistical nature of the excitation. This energy spread, $\sigma_{E_{beam}}$, is directly calculable from the machine parameters. Furthermore, as energy or phase oscillations translate directly into longitudinal oscillations, the beam energy spread can be checked by means of the

---

[5] In previous years four or eight equally spaced bunches were used.

experimental measurements of the longitudinal size of the beam spot together with the measured value of the oscillation frequency per turn, Q_s.

Without dispersion the centre-of-mass energy spread is given by

$$\sigma_{E_{\mathrm{CM}}} = \sqrt{2}\,\sigma_{E_{\mathrm{beam}}} \ . \tag{2.10}$$

The presence of opposite-sign vertical dispersion in 1995 reduces $\sigma_{E_{\mathrm{CM}}}$ by 2 %. Typically $\sigma_{E_{\mathrm{CM}}}$ is about 55 MeV for the 1993–1995 data, with a 2 % uncertainty.

A similar effect is caused by the fact that many LEP fills are combined to produce the experimental observables for a specific scan point. These fills are not at precisely the same energy but scatter by typically 10 MeV around the average, which needs to be added on top of the centre-of-mass energy spread.

Owing to the spread the measured observables, σ_f or $A_{\mathrm{FB}}^{\ell\ell}$, need to be corrected to a sharp energy before they are compared with theoretical parametrizations in the fit. In a lowest-order Taylor expansion the correction for the cross-section is given by

$$\Delta\sigma_f(E_{\mathrm{CM}}) = -\frac{1}{2}\frac{\mathrm{d}^2\sigma_f(E_{\mathrm{CM}})}{(\mathrm{d}E_{\mathrm{CM}})^2}\sigma_{E_{\mathrm{CM}}}^2 \ . \tag{2.11}$$

Ignoring this effect would cause a systematic reduction in the peak cross-section by about $-0.1\,\%$ and a 4 MeV increase of the Z^0 width, much larger than the experimental errors.

Further Effects. Now all effects which cause a substantial correction or uncertainty have been discussed. A few minor corrections remain, as follows.

- Positron beam energy. In general the resonant-depolarization measurement was made with the electron beam. In principle differences can arise since electrons and positrons have a different orbit (the "saw tooth"; see Fig. 2.4). To ensure consistency a few resonant depolarization measurements with the positron beam were made; these agreed within 0.3 MeV. Similarly, resonant-depolarization measurements of different bunches were consistent within errors.
- Calibration versus physics conditions. The tuning of the LEP optics is quite different for resonant-depolarization measurements and physics collisions, for example corrector dipole settings, solenoid compensation and betatron tunes usually need to be adjusted. Many dedicated depolarization experiments proved either that these effects are negligible or that their impact is consistent with the expectation within the resolution of the measurement.
- Static magnetic fields. Resonant depolarization measures directly the average energy in the ring; therefore static magnetic fields caused, for example, by the magnetization of the environment are of no concern. But it is interesting to note that even the contribution of the earth's magnetic field to the bending corresponds to 0.9 MeV in the beam energy.

Calibration Summary. All sources of energy variations are comprised in the LEP energy model, which parametrizes in a first stage the beam energy at a given point in time:

$$E_{\text{beam}}(t) = E_{\text{nom}}(\text{fill}) \left[1 + C_{\text{rise}}(t_{\text{day}}, t_{\text{fill}}) + C_{T\text{-dipole}}(t) \right.$$
$$\left. + C_{\text{tide}}(t) + C_{\text{orbit}}(\text{fill}) + C_{\text{other}}(t) \right]. \tag{2.12}$$

- $E_{\text{nom}}(\text{fill})$ is taken from the depolarization measurement of this fill, if it has been performed. Otherwise it is a properly weighted average of all the fills for that energy point.
- $C_{\text{rise}}(t_{\text{day}}, t_{\text{fill}})$ describes the dipole field change due to parasitic currents. It depends on the time of day (variations in the train schedule) and the time in the fill (saturation of the rise).
- $C_{T\text{-dipole}}(t)$ is the temperature correction for the dipole field averaged over the LEP ring.
- $C_{\text{tide}}(t)$ parametrizes the tidal forces.
- $C_{\text{orbit}}(\text{fill})$ accounts for the slow variation of the beam orbit due to hydrogeological deformations.
- $C_{\text{other}}(t)$ subsumes various other small terms, such as those due to the horizontal magnet correctors.

Then, in a second step, the interaction-point-specific corrections, depending on the RF configuration and the dispersion, are applied:

$$E_{\text{CM}}^{\text{IP}}(t) = 2E_{\text{beam}}(t) + \Delta E_{\text{RF}}(t) + \Delta E_{\text{disp}}(t). \tag{2.13}$$

Such a detailed model guarantees the experiments full flexibility in their data analysis; no bias arises when fills need to be discarded partially or as a whole owing to data quality requirements.

Table 2.3 lists the contributions of the individual effects to the systematic uncertainty for the 1993 and 1995 off-peak scan points. The dipole field rise causes the largest error; in particular for 1993, it dominates the overall uncertainty. The total error amounts to about 3.1 MeV in 1993 and 1.7 MeV in 1995. For a precise determination of the error on m_Z and Γ_Z a careful analysis of correlations between different energy points is required, since the latter is insensitive to fully correlated uncertainties while the former benefits from small correlations.

2.2.2 Luminosity Measurement

The luminosity determination makes use of small-angle e^+e^- Bhabha scattering as a reference process. Experimentally, $e^+e^- \to e^+e^-$ events at small angles are counted (N_e) and the normalization with the theoretically calculated cross-section σ_e^{th} yields the integrated luminosity

$$\int \mathcal{L}\, dt = \frac{N_e - N_{\text{backg}}}{\varepsilon_e\, \sigma_e^{\text{th}}}, \tag{2.14}$$

Table 2.3. Breakdown of errors (in MeV) on the centre-of-mass energy measurement at LEP. The numbers are for illustration only; the full determination is done with the complete correlation matrix

	-2 1993	0 1993	$+2$ 1993	0 1994	-2 1995	0 1995	$+2$ 1995
Normalization error	1.7	5.9	0.9	1.1	0.8	5.0	0.4
Resonant depolarization	0.5	0.5	0.5	0.5	0.5	0.5	0.5
Quadrupole corrections	0.2	0.2	0.2	0.0	0.0	0.0	0.0
Horizontal correctors	0.0	0.4	0.4	0.2	0.2	0.5	0.2
Tide amplitude	0.0	0.3	0.2	0.1	0.0	0.0	0.0
Tide phase	0.0	0.0	0.1	0.1	0.2	0.0	0.0
Ring temperature	0.1	0.4	0.4	0.2	0.4	0.3	0.4
B rise scatter + model	2.8	3.0	2.5	3.3	0.6	0.6	0.6
B rise temperature correction	0.6	0.3	0.6	0.5	1.0	1.0	1.1
Bending modulation	0.0	0.0	0.0	0.0	0.0	1.4	0.3
e^+ energy uncertainty	0.3	0.3	0.3	0.3	0.2	0.2	0.2
RF correction	0.5	0.5	0.5	0.6	0.7	0.7	0.7
Dispersion correction	0.4	0.4	0.4	0.7	0.3	0.3	0.3

where N_{backg} and ε_e account for the background and the efficiency of the event selection.

Bhabha scattering is ideally suited for this purpose. Firstly, it is dominated by t channel photon exchange and the cross-section can be computed very accurately in the framework of QED. γZ interference contributions are well below 1 % in the region of the Z^0 resonance. Secondly, it has a very clear experimental signature and the rate of luminosity events exceeds the rate of events from Z^0 decays when the angular region is appropriately defined.

The major challenge in the luminosity measurement is caused by the strong dependence of the small-angle Bhabha cross-section on the scattering angle

$$\frac{\mathrm{d}\sigma_{e^+e^-}}{\mathrm{d}\theta} \approx \frac{32\pi\alpha^2}{s}\frac{1}{\theta^3} . \tag{2.15}$$

In particular, the angle of the inner edge of the detector with respect to the beam axis, $\theta_{\min}$, needs to be known with high precision.

Before LEP, e^+e^- experiments typically reached an accuracy of 2–4 % for the luminosity and such a precision was also foreseen in the original specifications of the LEP detectors. Improved analysis techniques together with progress in theoretical calculations reduced the luminosity uncertainty to about 0.5 % in the early years of LEP data-taking (1989–1992). But it became apparent that the existing LEP luminosity detectors would prevent any substantial further improvement. At the same time tremendous progress was being made in the design and production of semiconductor-based tracking devices, which opened up a new regime of precision in position reconstruction. As a result all four LEP experiments decided to build "second generation" luminosity detectors with the goal of reducing the experimental uncertainty to 0.1 % or below in order to match the statistical precision of the recorded hadronic Z^0 decays. ALEPH [142], L3 [143] and OPAL [144] based the luminometer on sandwich calorimeters with a silicon layer for the position reconstruction. DELPHI [145] took a different approach using a precisely machined tungsten absorber to define $\theta_{\min}$.

In the following the main ingredients of the the luminosity measurement are briefly described.

- Position reconstruction. Typically, the new LEP luminosity detectors are placed 2.5 m away from the interaction point and their angular fiducial region extends from $\theta_{\min} = 30$ mrad to $\theta_{\max} = 50$ mrad, which corresponds to radii of about 7 cm and 15 cm from the beam axis. To reach 0.1 % precision requires a geometrical alignment within 25 μm, 100 μm and 1 mm for the inner and outer radii and the z position, respectively. The systematic bias in the position reconstruction of the outgoing electrons and positrons must be controlled at the same level. This reconstruction depends on the detector set-up: ALEPH and OPAL base it on the energy sharing across adjacent pads to exploit the precise mechanical construction. L3 makes use of a silicon tracker in front of the calorimeter and DELPHI relies on its tungsten shield, which translates the inner radius into an energy cut-off.

- Beam parameters. At the required level of precision, effects of the LEP beam become equally important. Any deviation of the interaction point from the nominal detector origin or tilts of the beam with respect to the detector axis change the acceptance. Similarly, the transverse and longitudinal beam size as well as the divergence of the beam can have sizeable effects. However, the sensitivity of the measurement to transverse beam variations can greatly be reduced by applying different fiducial cuts on the two sides of the detector, a "tight" region on one side and a "loose" region on the other, the latter typically allowing a 40 % wider range in θ. This way any transverse beam offset or tilt affects the luminosity only to second order. In addition, by switching the sides of the loose and tight cuts randomly, e.g. on an event-by-event basis, longitudinal offsets are also compensated to first order.[6]

- Background. The main background is caused by off-momentum electrons and positrons which are deflected by the focusing quadrupole magnets into the luminosity calorimeters. They can fake a Bhabha event when they hit the detector in coincidence on both sides. This background is largely

---

[6] In DELPHI the use of the tungsten shield to define $\theta_{\min}$ prohibits an alternate loose/tight cut.

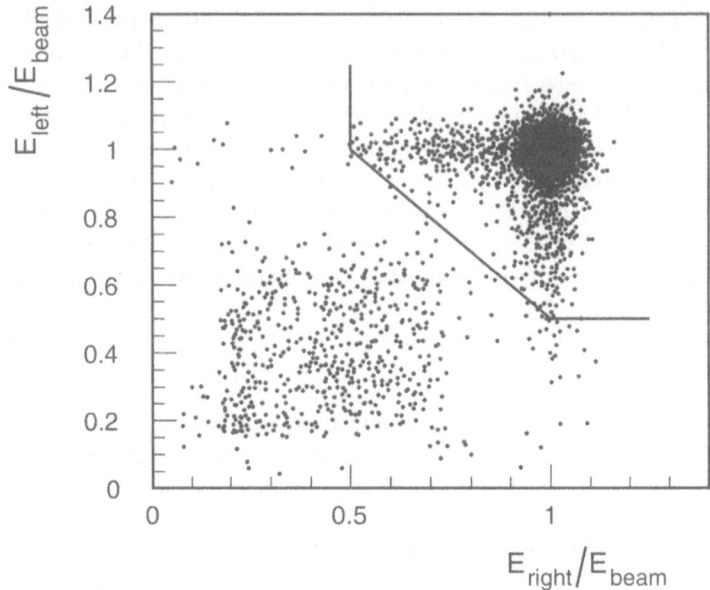

Fig. 2.8. Measured energy on the two sides of the calorimeter (OPAL [144]). Genuine Bhabha events deposit approximately the beam energy on both sides. Radiative events are responsible for the horizontal and vertical tails, while the background from off-momentum electrons is well isolated at much lower energies

rejected by cuts on the energy reconstructed in the two calorimeters (see Fig. 2.8). The systematics can be nicely controlled with the data by combining the observed rate of events with hits on one side only. This independent background measurement is important as the rate depends on the vacuum conditions in the beam pipe and varies with time. Typically, it contributes at the 10^{-3} level or below. Other sources such as two-photon reactions or $e^+e^- \to \gamma\gamma$ are negligibly small.

• Theoretical uncertainties. In parallel with the experimental improvements, much effort also went into the calculation of the theoretical cross-section and the development of Monte Carlo generators (see e.g. [148]) which allow a detailed mapping of the experimental cuts to the accepted phase space. The precision of the Bhabha cross-section calculation for the LEP luminosity detectors could be improved from about 1 % in 1989 to 0.11 % in 1995, by including higher-order QED corrections and a detailed evaluation of remaining uncertainties [146,105]. The dominant uncertainty is due to missing second-order subleading $\mathcal{O}(\alpha^2 L)$ corrections ($L = \log(-t/m_e^2)$). A further reduction of the theoretical uncertainty to 0.06 % was achieved owing to recent progress in the evaluation of this correction [147].

It is worth noting that theoretical uncertainties need to be considered for the design of the selection cuts, as it is favourable to develop an inclusive selection which imposes no stringent phase space limits for initial- or final-state radiation. Therefore, the use of tight and loose fiducial cuts in θ together with modest requirements on the minimal energy and maximal acollinearity help to minimize theoretical uncertainties.

In summary, the introduction of second-generation luminosity detectors provided a huge improvement, by almost an order of magnitude, in the experimental precision of the luminosity measurement. The uncertainty obtained ranges from 0.037 % for OPAL, 0.044 % for ALEPH and 0.078 % for L3 to 0.09 % for DELPHI. In general, all the above-mentioned sources – mechanical precision, inner-edge resolution, beam parameters – contribute at a similar level to the error. Only the DELPHI measurement, which is by construction more sensitive to longitudinal beam variations, is dominated by this single effect.

2.2.3 Event Selections

Owing to the clean operation conditions at the LEP collider and the high redundancy of the detectors, the four final-state categories – hadronic events, and electron, muon and tau pairs – are triggered with essentially 100 % efficiency and show very characteristic signatures. Background from non-Z^0 decays is largely suppressed owing to the large cross-section at the Z^0 resonance.

Figure 2.9 shows typical examples of the four event categories which illustrate the characteristic features. Hadronic events have large multiplicity in the tracking chambers and the calorimeters. $e^+e^- \to e^+e^-$ reactions show two back-to-back high-momentum tracks associated with large energy in the electromagnetic calorimeter. $\mu^+\mu^-$ pairs have a similar track pattern but the associated calorimeter energies are small and the muon chambers provide additional information. $e^+e^- \to \tau^+\tau^-$ events have the least unique signature as the τ leptons produced in the reaction decay after few millimetres and the observed pattern depends on the decay mode. An important feature is the missing energy carried away by the τ neutrino in the decay.

Owing to these clear signatures none of the selections depends strongly on the performance of the detectors in terms of the common benchmarks such as position, momentum and energy resolution, particle identification or vertex finding. Instead, the crucial elements are good hermeticity, homogeneous response, redundant components and stable operation.

In principle the analysis is straightforward: only a few, rather basic cuts are needed to obtain selections with high efficiency and purity. Applying the selection criteria to the corresponding signal and background Monte Carlo samples yields the necessary correction factors for the cross-section calculation.

The Monte Carlo generators commonly used to simulate e^+e^- reactions at LEP1 energies are JETSET [153] and HERWIG [154] for $e^+e^- \to q\bar{q}$,

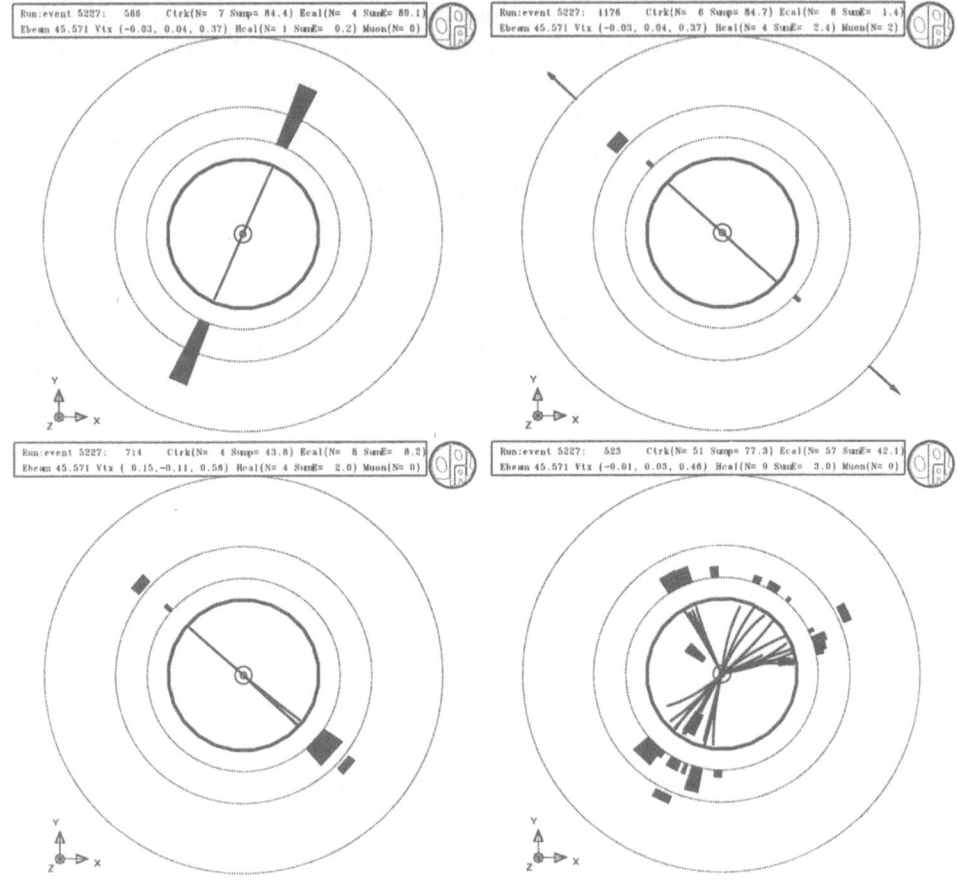

Fig. 2.9. Examples taken from the OPAL experiment for the four event categories. $Z^0 \to e^+e^-$ *(top left)*, $Z^0 \to \mu^+\mu^-$ *(top right)*, $Z^0 \to \tau^+\tau^-$ *(bottom left)*, $Z^0 \to$ hadrons *(bottom right)*

KORALZ [152] for $e^+e^- \to \mu^+\mu^-$ and $e^+e^- \to \tau^+\tau^-$ and BHAGEN [149], BABAMC [150] and BHWIDE [151] for $e^+e^- \to e^+e^-$.

The main challenge is to control the systematic uncertainties at the level of the high statistical precision of around 0.1 % as given by the large event sample at LEP1. Although the Monte Carlo simulations of event generation and detector response are able to reproduce the observed data properties in considerable detail, this is not sufficient to control systematic uncertainties below the per mille level. In order to reach such an accuracy the analyses, in general, use Monte Carlo simulations as a first iteration step only. Any model-dependent corrections are derived by making best use of the redundancy which is available from different detector components or inherent in the event.

Table 2.4. Experimental systematic uncertainties for luminosity, event selections and asymmetry of the LEP experiments for the 1994 data-taking [155]

	ALEPH	DELPHI	L3	OPAL
$\mathcal{L}_{\mathrm{exp}}$	0.044 %	0.09 %	0.078 %	0.033 %
σ_{had}	0.071 %	0.10 %	0.051 %	0.072 %
σ_e	0.16 %	0.52 %	0.23 %	0.16 %
σ_μ	0.09 %	0.26 %	0.31 %	0.12 %
σ_τ	0.18 %	0.60 %	0.65 %	0.48 %
A_{FB}^e	0.0012	0.0021	0.003	0.001
A_{FB}^μ	0.0005	0.0005	0.0008	0.001
A_{FB}^τ	0.0007	0.0020	0.003	0.0012

In the following the main elements of the selections are discussed. More details can be found in the publications and notes from the experiments [156–159]. The overall systematic uncertainties for the event selections and the asymmetries are summarized in Table 2.4.

Hadronic Events

About 87 % of the visible Z^0 decays are to quark–antiquark pairs, leading to hadronic final states. These events therefore have the potential to provide the most accurate determination of Z^0 properties, such as m_Z and Γ_Z. In general, hadronic events can be clearly distinguished from leptonic Z^0 decays and other background processes owing to their large particle multiplicity and high energy visible in the detector.

Inefficiencies arise mainly for events in which most of the final-state particles have trajectories with only a very small angle relative to the beam axis. The geometrical coverage of the detector is incomplete in these regions; this is inevitable because of the finite size of the beam pipe and the associated apparatus and implicit in the rapidly falling efficiency of the tracking detectors. Therefore only a fraction of the particles from such events is registered in the detector. The more a hadronic event approaches a narrow two-jet-like topology the more likely it is to be rejected by the selection.

Two effects are crucial for controlling the systematic uncertainty of the hadronic selection. Firstly, the stability of the detector operation and the quality of the simulation are most important at low angles, as in this region the cuts act almost in the centre of the distributions and small deficiencies can potentially cause large effects. The quality of the data and simulation in the more central detector regions is less important as the cuts are applied in the tails. Secondly, the rate of lost events depends on the details of the event topology, particle composition and spectra, which are directly related to the models used in the simulation of the hadronization phase in the Monte Carlo event generators. The non-perturbative hadronization is based on phenomenological models which depend on empirically determined parameters. The selection efficiency can vary significantly when these parameters are changed or different models are used.

The main backgrounds originate from $\tau^+\tau^-$ events and two-photon reactions, in which two quasi-real photons radiated from the incoming beams collide and produce a quark–antiquark pair. Both can be kept at a low level, typically below 0.5 %.

The selections applied by the LEP experiments all employ the same characteristic features. The multiparticle structure of hadronic events is used to discriminate against lepton pairs and the energy and momentum balance is used to minimize background from two-photon events. But each experiment has specific preferences as to what subdetectors are used to construct the cuts. Evidently, the trade-off between extending the angular range as far as possible to maximize the acceptance and limiting this range in order to remain in the efficient and well-understood regions of the detector results in several quite different strategies.

- ALEPH uses two different selections, one based on the tracking in the TPC alone, the other using the calorimeter system for the energy-related cuts. The former provides an efficiency of 97.5 %, the latter 99.1 %, owing to the more complete coverage of the solid angle by the calorimeters. The results of the two selections are averaged; thereby uncorrelated systematic effects are reduced.
- DELPHI's analysis is based mainly on charged tracks, which are restricted to the angular range $20° < \theta < 160°$. A further cut on the maximal energy in the electromagnetic calorimeter suppresses background from electron pairs. The efficiency is relatively small, with a value 95.2 %, owing to the limited angular range for tracks.
- L3 uses the electromagnetic and hadronic calorimeter information. It achieves a high efficiency of 99.2 %, taking advantage of the large coverage in θ of the hadronic calorimeter.
- OPAL constructs the cut variables on the basis of a combination of charged tracks and clusters in the electromagnetic and forward calorimeters, aiming also to maximize the sensitive solid angle. This way the acceptance is extended to 99.5 %.

Different approaches are also taken in the evaluation of systematic errors. ALEPH controls hadronization uncertainties in the TPC selection by rotating events well measured in the central region into the forward region, accounting for the efficiency change at small angles. Doing this in parallel for data and Monte Carlo events gives an estimate of possible hadronization modelling deficiencies.

A similar method is employed by OPAL; this collaboration emulates in the central region the layout and response of the calorimeters in the forward direction, in both data and Monte Carlo events. The rate of lost events in this "barrel hole" can be directly measured with the data. The Monte Carlo simulation is only used to extrapolate the inefficiency to the forward region. Thereby, even large variations of the hadronization parameters or models have negligible effects on the calculated acceptance. However, a stronger dependence on the quality of the detector simulation in the central detector regions is introduced by this technique. As OPAL's selection relies primarily on the reliable operation and simulation of the electromagnetic calorimeters, extensive studies were performed using isolated hadrons and photons to verify both the homogeneity of the detector response and the quality of the simulation. Isolated hadrons and photons avoid the principal difficulty of disentangling hadronization effects from detector simulation deficiencies; they can be used to assess locally the detector response.

The background from τ pairs as such is small, with a value of 0.2–0.5 %. However, owing to the factor of 20.7 between hadronic and leptonic cross-sections this translates into a 4–10 % acceptance for τ pairs, so a substantial fraction of $\tau^+\tau^-$ events is included in the hadronic samples. The selection cuts are applied in the steeply falling upper end of the $\tau^+\tau^-$ multiplicity spectrum; therefore any small change in the shape of this spectrum can lead to rather large effects. One potential source affecting the multiplicity is photons converting to e^+e^- pairs in the detector material. Photons are often produced in τ pair events, originating from τ decay products such as π^0 or from final-state radiation, and the simulation of conversion is rather delicate as it requires a very detailed mapping of the various detector materials in the simulation. Therefore, a good understanding of photon conversion is particularly important for a precise estimation of the τ background.

The other "large" background source is two-photon reactions, typically 0.05–0.2 % at the peak of the resonance. These reactions are not mediated by the Z^0 and the cross-section varies little around the Z^0 resonance. Therefore the fractional background depends on the centre-of-mass energy, and the uncertainty affects not only the absolute normalization but also the Z^0 width. The main difficulty in estimating this background is theoretical uncertainties in the physics modelling of two-photon reactions, which affects both the overall cross-section and the spectrum of the variables (momentum sum, calorimeter energy, energy balance) used for the discrimination.

DELPHI and L3 fit two-photon Monte Carlo simulations to the observed data distributions and derive systematic errors from variations of the fit range and Monte Carlo models. ALEPH and OPAL employ the specific features of two-photon events, low energy and large momentum imbalance, together with their fairly constant cross-section in the vicinity of the Z^0 resonance to estimate the background directly from the data.

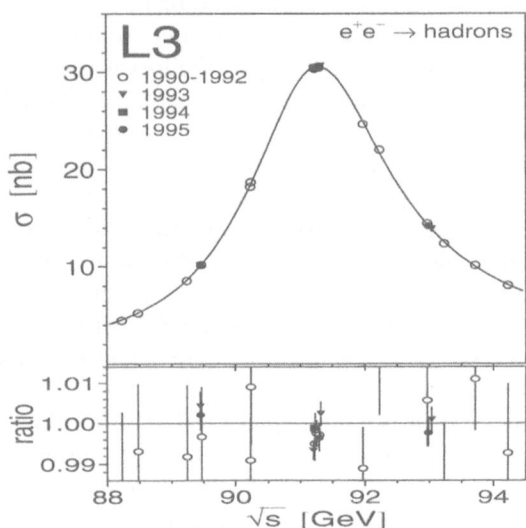

Fig. 2.10. Measurements of the hadronic cross-section by the L3 collaboration [158]

Further background sources are e^+e^- events,[7] cosmic rays and machine background originating from interactions with gas particles in the beam pipe or the beam pipe wall. In general these are very small, 1×10^{-4} or less.

In addition to the systematic studies performed for the determination of the efficiency and background, the high precision required makes it necessary to investigate in detail the operational stability of the detector and the reliability of the reconstruction chain. Given the complexity of the experimental apparatus and the multitude of steps needed to go from the collision in the collider to the data files in the offline analysis farm it is far from trivial to ensure that the event sample entering the analysis is consistent with the events produced below the 1×10^{-4} level over many years.

Overall, the systematic uncertainties of the hadron event selection quoted by the LEP experiments range between 0.05 % and 0.09 % (Table 2.4), balancing approximately the statistical error.

Figure 2.10 shows an example of the measured cross-sections for hadronic events around the Z^0 resonance.

Leptonic Events

Leptonic events can be rather cleanly separated from hadronic events using cuts based on the particle multiplicity in the event. The distinction between

---

[7] Occasionally an electron initiates a particle shower already in the beam-pipe, which can result in a large particle multiplicity.

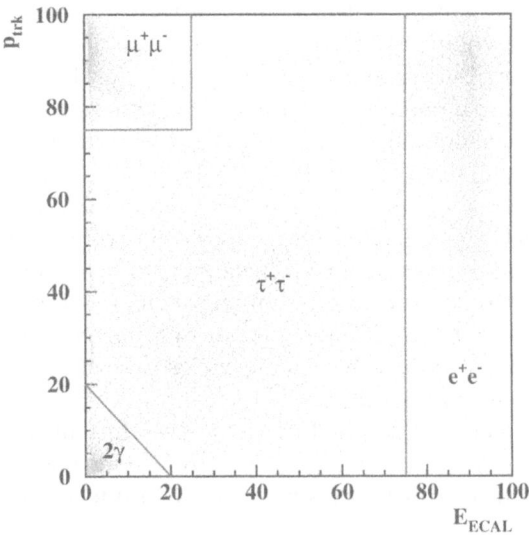

Fig. 2.11. Separation of e^+e^-, $\mu^+\mu^-$ and $\tau^+\tau^-$ final states with charged momentum ($p_{\rm trk}$) and electromagnetic energy ($E_{\rm ECAL}$)

the three leptonic decay channels can be accomplished with the charged momentum and electromagnetic energy as illustrated in Fig. 2.11.

In contrast to hadronic events the sensitive angular range of the tracking detectors implies a cut-off in the polar angle for leptonic events; the selections are restricted to a predefined region in θ. Moreover, modelling uncertainties are in general negligible for the lepton pair selections since hadronization effects are absent (e^+e^- and $\mu^+\mu^-$) or very small ($\tau^+\tau^-$) and the Monte Carlo generators include higher-order QED corrections. On the other hand the sensitivity to local inhomogeneities in the detector response can be large as only few particles are created. Substantial information needed for the selection can be lost when particles traverse areas with poor resolution, dead zones or gaps in the detector.

In cases where both τ leptons decay to either electrons or muons a small but unavoidable ambiguity exists in separating these from genuine electron or muon pairs. It is advantageous to use common separation cuts in the individual selections which define the border between the genuine e^+e^- or $\mu^+\mu^-$ on one side and $\tau^+\tau^-$ events on the other. This way such ambiguities in the classification of the individual lepton species largely cancel when lepton universality is assumed in the interpretation.

Electron Pairs. The analysis of Z^0 decays into e^+e^- pairs is complicated by contributions from t channel scattering. In contrast to the luminosity measurement, which exploits the high rate of this process at low angles and its insensitivity to electroweak corrections, the e^+e^- event selection aims to reduce these contributions in order to retain a good sensitivity to Z^0 properties.

As the t channel contribution rises steeply with smaller scattering angles the polar angle of accepted events is commonly restricted to $|\cos\theta| < 0.7$.[8] This not only reduces the overall contribution from t channel scattering but also the uncertainty which arises from the precision of the θ determination.

The selections typically reach high efficiencies of $\geq 99\,\%$ within the accepted angular range. Backgrounds are low ($< 0.5\,\%$), mainly τ pairs and $e^+e^- \to \gamma\gamma$ reactions. The inherent redundancy of e^+e^- events – two high-momentum back-to-back tracks with associated energy in the electromagnetic calorimeter – gives powerful handles to control the efficiency and background directly with data. Modifying the cuts, e.g. by applying stringent criteria only on one side of the event, only for track momenta or only for calorimeter energies, allows detailed checks on possible inefficiencies or background contributions not accounted for by the simulation.

One of the largest systematic errors is caused by the uncertainty of the θ determination at the edge of the accepted angular range. It is controlled by comparing different methods of measuring θ: track or calorimeter information, electron or positron angles.

The overall systematic uncertainties of the e^+e^- selections range from $0.14\,\%$ to $0.30\,\%$ (Table 2.4).

Muon Pairs. Like e^+e^- pairs the $Z^0 \to \mu^+\mu^-$ process has a very clear and unique signature. As $e^+e^- \to \mu^+\mu^-$ is a pure s channel annihilation process the accepted angular range is limited only by the detector inefficiency at low angles; therefore the range can be extended to larger angles ($|\cos\theta| < 0.90$ or 0.95). Since muons usually penetrate all detector components the identification can make use of the hadron calorimeter and the muon chambers as well as the tracking chamber and the electromagnetic calorimeter, which gives a high redundancy and direct constraints on efficiency and background from the data.

The main background is $\tau^+\tau^-$ pairs, typically at the $1\,\%$ level. Muons produced in cosmic rays can mimic a $\mu^+\mu^-$ pair. Information from the time-of-flight system and the primary vertex greatly reduces this background and can also be used to estimate the residual contamination.

The total systematic uncertainties are in the range from $0.10\,\%$ to $0.30\,\%$ (Table 2.4). One of the largest contributions is the limited accuracy with which the angle of the edge of the geometrical acceptance is known.

Tau Pairs. Compared with e^+e^- and $\mu^+\mu^-$, the selection of $\tau^+\tau^-$ events is less clear as it is obscured by the secondary decays of the τ. The creation of neutrinos in the decay causes a broad energy spectrum of the observed particles and the hadronic decay modes can result in large particle multiplicities. The selection needs to suppress four major background sources, each of these affecting a different region in phase space of the genuine $\tau^+\tau^-$ properties. On the "high-energy" side e^+e^- and $\mu^+\mu^-$ events need to be suppressed, which

---

[8] ALEPH uses an asymmetric range $-0.9 < \cos\theta < 0.7$.

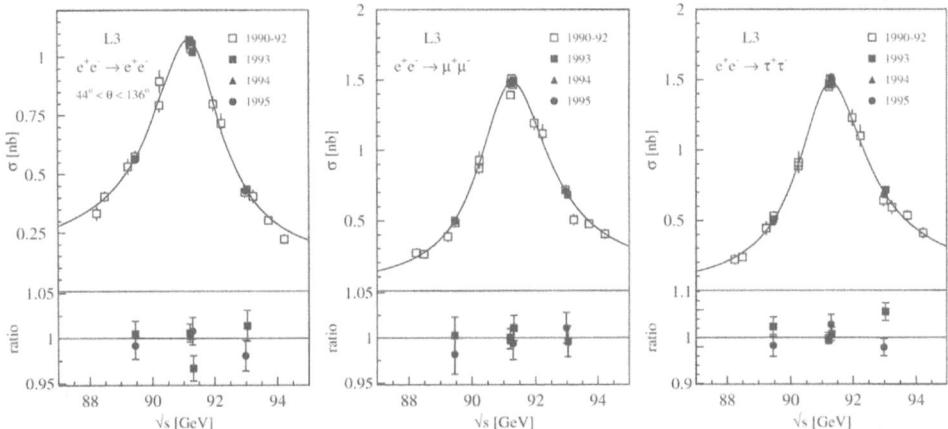

Fig. 2.12. Measurements of leptonic cross-sections by the L3 collaboration [158]. Note the different shape for σ_e (*left*) due to t channel contributions

is usually achieved with cuts on the charged momentum and electromagnetic energy, while at "low energies" two-photon events creating low-energy lepton pairs must be rejected. And finally, hadronic events can extend down to rather low particle multiplicities overlapping with $\tau^+\tau^-$ "high-multiplicity" events. Such hadronic events originate from the extreme tails of the fragmentation spectrum, and uncertainties in the Monte Carlo modelling of the hadronization process potentially give rather large effects.

Therefore the τ selection is considerably less efficient and pure when compared with the two other lepton pair selections; the acceptance is around 85–90 % within the angular range $|\cos\theta| < 0.9$ and background contributes at the 2–3 % level in total.

In principle the performance of the selection could be improved by applying more complicated cuts based on the specific event topologies or on more sophisticated particle identification capabilities of the detectors. For instance, it is rather unlikely that both τ leptons decay into the same lepton, while the background from lepton pairs or leptonic two-photon events always consists of two identical leptons in the final state. Similarly, hadronic events with low multiplicity mostly do not have the typical τ pattern of charged multiplicity – one or three charged tracks in each hemisphere – and they are also very unlikely to produce an isolated electron or muon in a hemisphere.

However, in general it is more difficult to control systematic errors when such complicated cuts are used. It is more advantageous to design a simple selection and use the advanced identification techniques to construct control samples which allow sensitive tests and corrections of the simulation. This way the overall systematic errors can be reduced nearly to the level reached in the selection of e^+e^- and $\mu^+\mu^-$ pairs, 0.2–0.6 % (Table 2.4). There is no clear dominating contribution; many effects add together with similar size.

Figure 2.12 shows an example of the measured cross-sections for leptonic events around the Z^0 resonance.

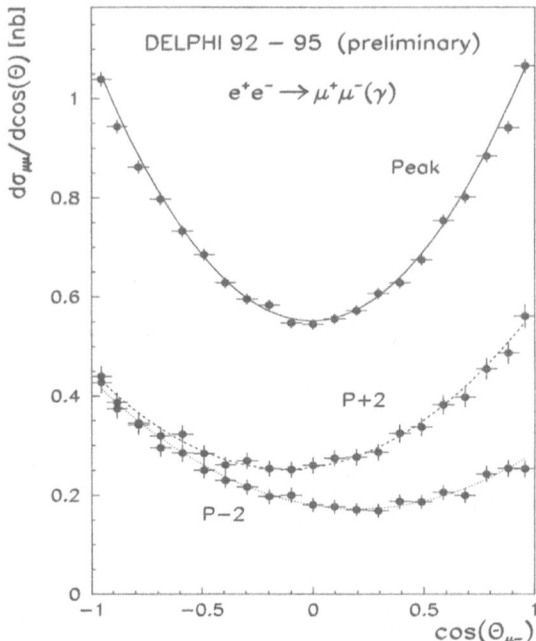

Fig. 2.13. Differential cross-section for $\mu^+\mu^-$ measured by the DELPHI collaboration [157]

Leptonic Forward–Backward Asymmetries

The measurement of the leptonic forward–backward charge asymmetry is based essentially on the same selections as used for the cross-section measurement. One difference is that often an additional cut on the acollinearity angle[9] is imposed, which suppresses events with large initial-state radiation. In order to make optimal use of the data statistics it is advantageous to determine the asymmetry for $\mu^+\mu^-$ and $\tau^+\tau^-$ in a maximum-likelihood fit to the angular distribution (Fig. 2.13)

$$\frac{d\sigma}{d\cos\theta} = \text{const}\left(1 + \cos^2\theta + \frac{8}{3}A_{FB}\cos\theta\right). \tag{2.16}$$

e^+e^- events have a more complicated angular distribution, owing to the t channel contribution. Therefore the asymmetry is determined from a simple counting method:

$$A_{FB} = \frac{N_{\theta<\pi/2} - N_{\theta>\pi/2}}{N_{\theta<\pi/2} + N_{\theta>\pi/2}}, \tag{2.17}$$

which is also used as a systematic check in the other channels.

---

[9] Acollinearity is defined as 180° minus the track opening angle.

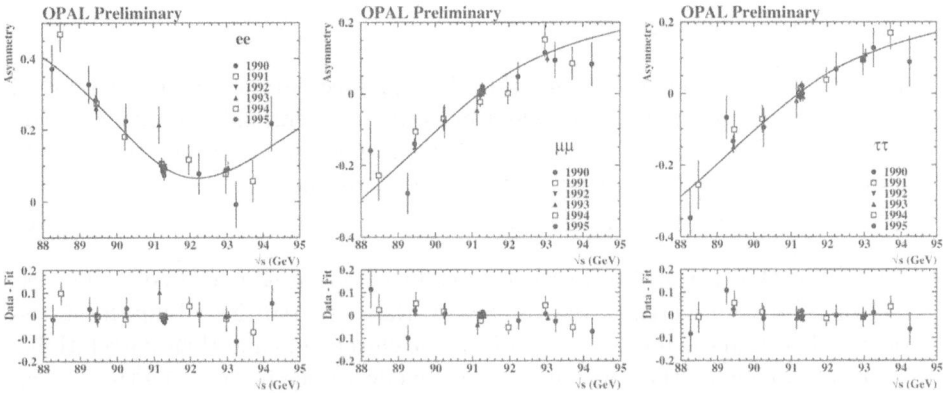

Fig. 2.14. Measurements of leptonic asymmetries by the OPAL collaboration [159]. Note the different shape for A_{FB}^e (*left*) due to t channel contributions

As the asymmetry is a ratio of cross-sections most systematic errors of the event selections cancel, except for the precision of the polar-angle determination and background uncertainties when they differ in their angular distribution. For the former, methods are applied to control the systematics that are similar to those used in the event selection for the cross-section: the use of different detector components for the θ measurement and using the positive or negative lepton to define θ. For the $\tau^+\tau^-$ asymmetry the e^+e^- background is particularly important as its angular distribution is very different.

On the other hand, effects which cause an asymmetric efficiency, a bias in the measurement of the polar angle θ or a charge misidentification cancel to first order for the cross-section but directly enter the asymmetry.

The charge is determined from the curvature of charged tracks. The misidentification probability can be directly measured from the data using the rate of events with the same charge in both hemispheres, which is typically at the 1 % level or below. Small differences between forward and backward efficiencies can be introduced by the properties of the physical reaction: e^+e^- events in the forward direction have a different rate and spectrum of photon radiation compared with backward events, since the t channel contributes much more strongly in the forward direction. For τ pairs the polarization and polarization asymmetry (Sect. 2.3.1) can also give rise to slight asymmetries in the acceptance. The energy spectrum of the decay products is different for left-handed and right-handed τ leptons and therefore a different fraction of events passes the selection cuts.

Figure 2.14 shows an example of the measured leptonic forward–backward asymmetries.

2.2.4 Lineshape Fit

The LEP experiments determine the Z^0 resonance parameters in a combined fit to the measured hadronic and leptonic cross-sections and leptonic forward–backward asymmetries. In the fit a set of parameters which describe the Z^0 properties is extracted in a χ^2 minimization procedure:

$$\chi^2 = \Delta^{\mathrm{T}} \otimes C^{-1} \otimes \Delta \,, \tag{2.18}$$

where Δ is the vector of residuals between the measured and predicted cross-sections and asymmetries and C is the covariance matrix describing the statistical and systematic uncertainties of the measurements and their correlations.

In the following the construction of the covariance matrix, the theoretical calculation of cross-sections and asymmetries and the choice of the lineshape parameters are briefly described.

Covariance Matrix

In general, the LEP experiments split their data samples for the fits into many subsamples, each comprising data from LEP fills of a restricted centre-of-mass energy range and a specific data taking period with similar detector configuration and performance. Typically, this results in about 30 values for each final state or about 200 cross-section and asymmetry measurements in total.

It is a non-trivial task to construct the corresponding 200×200 covariance matrix which maps in detail the correlations of the experimental errors for the 200 cross-section and asymmetry points.

Systematic uncertainties in the selections cause mainly correlations within the points of a specific final state. Complications arise from the different dependences on the centre-of-mass energy of the various effects and from changes of the selection criteria and the detector set-up over time. Further inter-species correlations also need to be taken into account.

Luminosity errors act simultaneously on all cross-section values of a specific period. As for the selection systematics, the energy dependence and the time evolution of the correlations needs to be considered.

Uncertainties in the beam energy and the beam energy spread are incorporated by translating the energy error into the corresponding cross-section or asymmetry error using the derivative of the observable with respect to the energy

$$C_{i,j} = \frac{\mathrm{d}\,\sigma(E_i)}{\mathrm{d}\,E} C_{i,j}^{\mathrm{ECM}} \frac{\mathrm{d}\,\sigma(E_j)}{\mathrm{d}\,E} \,, \tag{2.19}$$

where $C_{i,j}^{\mathrm{ECM}}$ denotes the LEP energy error matrix.

Theoretical Calculation

The properties of the Z^0 resonance, i.e. the parameters of a Breit–Wigner resonance (mass, width and peak height), can in principle be obtained by fitting a Breit–Wigner curve to the measured cross-sections. However, such a fit requires the unfolding of several effects such as initial-state radiation, contributions from photon exchange and other higher-order corrections in order to determine the pure Z^0 properties in terms of Breit–Wigner parameters.

Several theory groups provide semi-analytical computer programs for such calculations. The most commonly used among the LEP experiments are the ZFITTER [101] and TOPAZ [102] packages.

It should be pointed out that this need for unfolding inevitably prevents a rigorous model-independent parametrization of the Z^0 lineshape. Any parametrization depends on the specific treatment of the initial-state radiation and initial–final-state interference corrections, as this directly affects the unfolding. Depending on the parametrization, further dependences on the Standard Model predictions enter; this is discussed below.

For the e^+e^- final state further complications arise from the presence of t channel diagrams. In general, ZFITTER is used only for the s channel hard scattering process. Contributions from the t channel and s/t interference are calculated separately using ALIBABA [160] or TOPAZ [102] and then added to the pure s channel predictions for the fit. The theoretical uncertainties in these corrections are at the 0.1–0.2 % level for the cross-sections. These also need to be taken into account in the construction of the covariance matrix.

Model-Independent Parameters

In order to allow a transparent combination, the four LEP collaborations agreed to present their results for the lineshape fit in terms of nine parameters:

- mass and width of the Z^0 boson (m_Z, Γ_Z; see (1.133))
- hadronic pole cross-section ($\sigma^0_{\rm had}$; see (1.133))
- three ratios of hadronic partial width to the leptonic partial widths ($R_e = \Gamma_{\rm had}/\Gamma_e$, $R_\mu = \Gamma_{\rm had}/\Gamma_\mu$, $R_\tau = \Gamma_{\rm had}/\Gamma_\tau$; see (1.149))
- three leptonic pole asymmetries ($A^{0,e}_{\rm FB}$, $A^{0,\mu}_{\rm FB}$, $A^{0,\tau}_{\rm FB}$; see (1.165)).

If lepton universality is imposed, i.e. the couplings of the Z^0 are independent of the flavour of the charged lepton, only five parameters are used for the fit. The three partial-width ratios and asymmetries combine into a universal[10] leptonic partial width, $R_\ell = \Gamma_{\rm had}/\Gamma_\ell$, and pole asymmetry, $A^{0,\ell}_{\rm FB}$.

The theoretical predictions for the cross-sections and asymmetries depending on these parameters are calculated for each final state at each energy point.

---

[10] The different masses of e, μ and τ cause differences in the partial widths, which are only numerically significant for τ (1.149). The universal R_ℓ refers to massless leptons and the corresponding corrections are applied in the fit.

The specific choice of this set of nine or five parameters has the advantage of largely disentangling experimental systematic errors affecting the parameters and minimizing the correlations. Uncertainties in the absolute LEP energy scale concern essentially only m_Z, while relative energy errors mainly enter the precision of Γ_Z. Similarly, uncertainties in the luminosity calculation are absorbed in σ^0_{had} and cancel for all other parameters. Such a separation of experimental effects does not happen for other parametrizations, such as one in terms of partial widths ($\Gamma_{\mathrm{had}}, \Gamma_\ell$ instead of $\sigma^0_{\mathrm{had}}, R_\ell$ or effective couplings).

The label "model independent" ascribed to these parameters refers only to the pure Z^0 exchange couplings. The form and size of the Standard Model couplings are used for the contributions from γ exchange, γZ interference and the effects from initial-state radiation.

To reduce this model dependence some experiments have used more general approaches. OPAL performed a fit with six additional terms which parametrizes the γZ interference term for cross-section and asymmetry of the three leptonic final states [159]. L3 [158] applied the S-matrix approach, a complementary theoretical ansatz to describe the hard scattering process (Sect. 1.4.2, (1.137), [82]).

2.2.5 Combination of the Lineshape and Asymmetry Measurements

Ideally, the fit parameters would be determined in a global lineshape fit to the cross-section and asymmetry measurements of all four LEP experiments, which would allow a detailed account to be taken of all correlated systematic uncertainties both within one experiment and across experiments. However, this would require a common structure for all sources of correlations which contribute to the full error matrix of the combined cross-section and asymmetry data set, which consists of some 800 points. Although the underlying effects on the systematics are similar for each experiment, the methods of mapping these into the covariance matrix evolved independently and to streamline the individual techniques into a common standard would require a huge effort, not only in technical terms.

Instead, the LEP collaborations combine the results on the level of the model-independent parameters, taking into account the full parameter covariance matrix determined by the experiments and also the correlated uncertainties in the parameters among the experiments.

Common errors arise from the uncertainties in the LEP centre-of-mass energy and from the use of Monte Carlo event generators. For the latter in particular, the theoretical error in the luminosity of 0.06 % is crucial. Similar common uncertainties exist for the t channel correction of the e^+e^- final state. Possible common systematics could also arise from the hadronization uncertainties in the $e^+e^- \to q\bar{q}$ event generators. However, each experiment uses its own tuning of model parameters and the methods to assess the error are quite different; therefore the correlations among the experiments are negligibly small.

For the combination, the individual covariance matrices of the parameter fits of each experiment ($C_{\text{exp}}^{\text{par}}$) are combined with the common covariance matrices ($C_{\text{com}}^{\text{par}} = C_{\text{ECM}}^{\text{par}} + C_{\text{lumi}}^{\text{par}} + C_{t\,\text{ch}}^{\text{par}}$) to form the full LEP error matrix, as illustrated in Table 2.5. In case of the luminosity and t channel uncertainties

Table 2.5. Schematic representation of the covariance matrix used for the combination of the lineshape and asymmetry results of the LEP experiments

	ALEPH	DELPHI	L3	OPAL
ALEPH	$C_{\text{ALEPH}}^{\text{par}}$	$C_{\text{com}}^{\text{par}}$	$C_{\text{com}}^{\text{par}}$	$C_{\text{com}}^{\text{par}}$
DELPHI	$C_{\text{com}}^{\text{par}}$	$C_{\text{DELPHI}}^{\text{par}}$	$C_{\text{com}}^{\text{par}}$	$C_{\text{com}}^{\text{par}}$
L3	$C_{\text{com}}^{\text{par}}$	$C_{\text{com}}^{\text{par}}$	$C_{\text{L3}}^{\text{par}}$	$C_{\text{com}}^{\text{par}}$
OPAL	$C_{\text{com}}^{\text{par}}$	$C_{\text{com}}^{\text{par}}$	$C_{\text{com}}^{\text{par}}$	$C_{\text{OPAL}}^{\text{par}}$

it is straightforward to translate them into errors of the model-independent parameters and construct the $C_{\text{lumi}}^{\text{par}}$ and $C_{t\,\text{ch}}^{\text{par}}$ matrices: the luminosity affects only σ_{had}^0 while the t channel contributes only to R_e and $A_{\text{FB}}^{0,e}$. Furthermore, both are fully correlated and essentially the same for all data-taking periods.

A more sophisticated procedure is needed for the LEP energy errors as they vary throughout the LEP years and depend on the energy points. This prevents a direct translation of the LEP energy error matrix into the corresponding errors on the model-independent parameters. Instead, the latter errors are determined by repeating the full lineshape fit to a single experiment, applying different scale factors (f_{exp}) to the experimental and the energy error matrix. For example, the experimental errors may be halved ($f_{\text{exp}} = 0.5$) and the energy errors remain unchanged. This fit yields a parameter covariance matrix $C_{1/2}^{\text{par}}$ which can be written as

$$C_{1/2}^{\text{par}} = C_{\text{exp},1/2}^{\text{par}}/4 + C_{\text{ECM}}^{\text{par}} \,, \tag{2.20}$$

where $C_{\text{ECM}}^{\text{par}}$ represents the effective parameter error matrix for the LEP energy uncertainties and $C_{\text{exp}}^{\text{par}}$ all other uncertainties. Similarly, the resulting matrix for the standard fit can be written as

$$C_1^{\text{par}} = C_{\text{exp},1}^{\text{par}} + C_{\text{ECM}}^{\text{par}} \,. \tag{2.21}$$

By combining these two equations the effective parameter error matrix for the energy can be calculated:

$$C_{\text{ECM}}^{\text{par}} = \frac{4}{3} C_{1/2}^{\text{par}} - \frac{1}{3} C_1^{\text{par}} \,. \tag{2.22}$$

This procedure was tested using different scale factors for the experimental errors. It was found that in order to obtain a correct estimate for the

Table 2.6. Results and correlations of the lineshape and asymmetry combination fit for five parameters under the assumption of lepton universality [155,161]

		m_Z	Γ_Z	σ_{had}^0	R_ℓ	$A_{\text{FB}}^{0,\ell}$
m_Z (GeV)	91.1871 ± 0.0021	1.00	-0.01	-0.05	-0.03	0.05
Γ_Z (GeV)	2.4944 ± 0.0024		1.00	-0.28	0.00	0.00
σ_{had}^0 (nb)	41.544 ± 0.037			1.00	0.19	0.01
R_ℓ	20.768 ± 0.024				1.00	-0.05
$A_{\text{FB}}^{0,\ell}$	0.01701 ± 0.00095					1.00

energy error matrix $C_{\text{ECM}}^{\text{par}}$ the value of the scale factor $f_{\exp}$ must be close to 1 (e.g. $f_{\exp} = 0.95$). Otherwise a bias is introduced, leading to an over- or underestimate of the energy error matrix $C_{\text{ECM}}^{\text{par}}$.

2.2.6 Lineshape Results

The result of the combination is given in Table 2.6 for the five parameters. The $\chi^2/\text{d.o.f.}$ [11] of the fit is $37/(36 - 5)$, which indicates fair consistency of the individual measurements. The largest correlations exist between Γ_Z and σ_{had}^0, with a value of -28%, and between σ_{had}^0 and R_ℓ, with a value of 19%. The other correlations are very small. The results of the individual

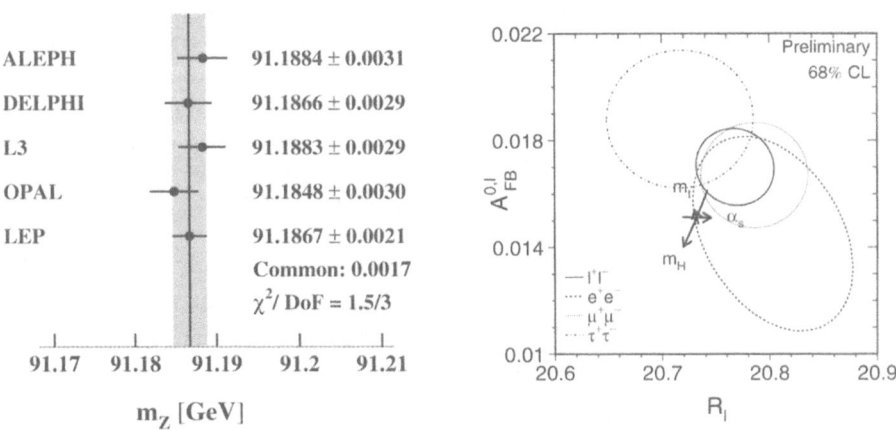

Fig. 2.15. *Left*: the Z^0 mass measured by the LEP experiments. *Right*: fit contours (68 % probability) of R_ℓ versus $A_{\text{FB}}^{0,\ell}$. Shown are the results for the nine-parameter fit, where the couplings of the three lepton species are treated independently, and the combined result assuming lepton universality. For ease of comparison the results for the τ are corrected to the massless case

---

[11] $\chi^2/\text{d.o.f.}$ is the χ^2 of the fit per degree of freedom.

experiments for the five parameters are shown in Figs. 2.15–2.17. In each case an approximate χ^2 of the average is given which takes the common part of the uncertainties into account. Except for m_Z, which is used as an input

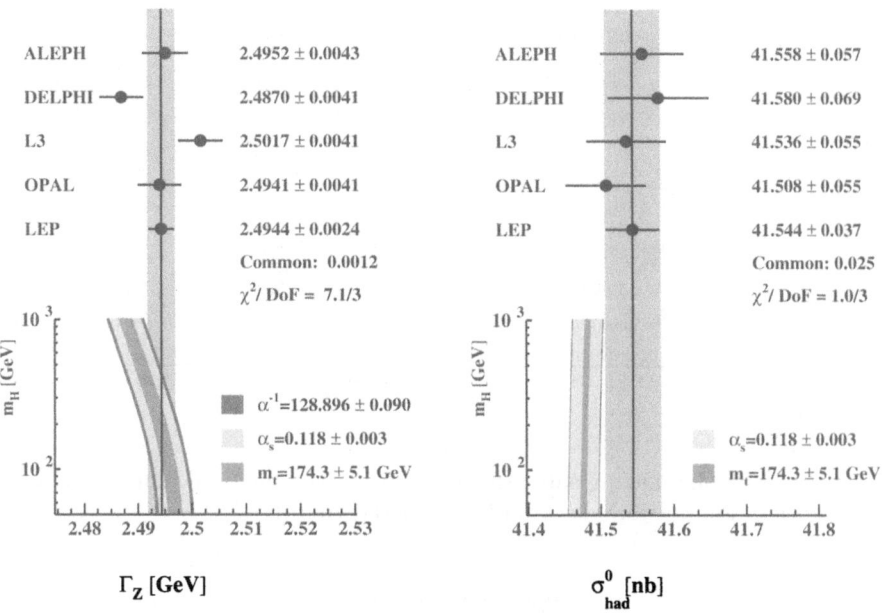

Fig. 2.16. LEP lineshape results: the Z^0 width Γ_Z (*left*) and the hadronic pole cross-section σ^0_{had} (*right*). The *hatched areas* of the Standard Model predictions indicate variations of the parameters m_t, α_s and $\alpha(m_Z^2)$

parameter to the Standard Model owing to its high precision, the figures also show the prediction of the Standard Model for each observable as a function of the mass of the Higgs boson m_H, and the variation with the top quark mass m_t and with the strong and electromagnetic coupling constants α_s and $\alpha(m_Z^2)$.

The combined LEP experiments measure m_Z with an extremely high accuracy of 23 ppm. Most of the uncertainty, namely 1.7 MeV or 80 % of the total error, is contributed by the LEP energy calibration. For the Z^0 width Γ_Z, the uncertainty is dominated by the statistical uncertainties in the measured hadronic cross-sections. The LEP calibration contributes about 1.2 MeV or 50 %. As shown in Fig. 2.16, Γ_Z is rather sensitive to m_H, but is also sensitive to variations of m_t and α_s within their experimental bounds. In contrast, σ^0_{had} depends mainly on α_s and is largely independent of the other Standard Model parameters. The uncertainty is largely dominated by the theoretical uncertainty of the luminosity measurement, which contributes about 0.025 nb or 70 %. The ratio of the hadronic to the leptonic Z^0 decay

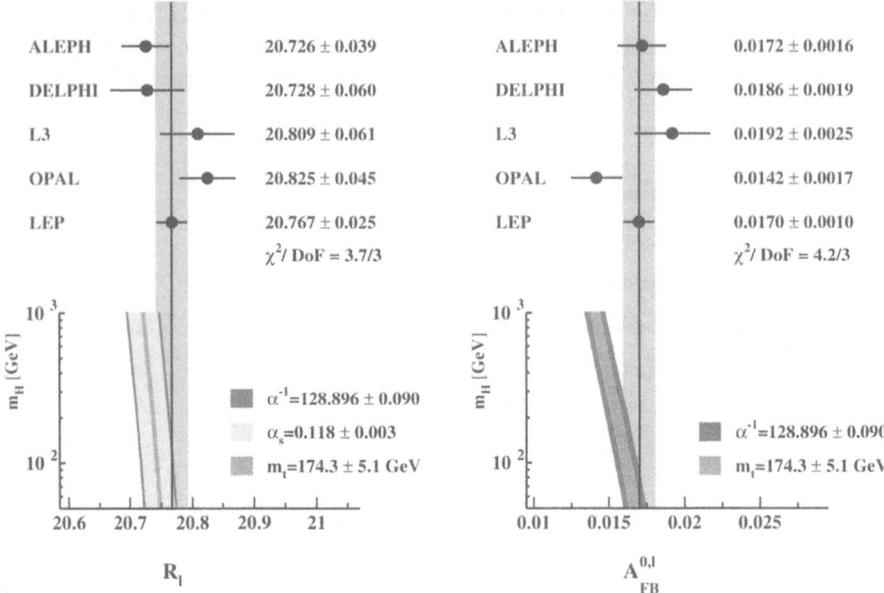

Fig. 2.17. LEP lineshape results: the hadronic to leptonic branching ratio, $R_\ell = \Gamma_{\mathrm{had}}/\Gamma_\ell$, (*left*) and forward–backward asymmetry, $A_{\mathrm{FB}}^{0,\ell}$, (*right*)

width, R_ℓ (Fig. 2.17), also depends mainly on the strong coupling constant α_{s}. Its error is composed about equally of the leptonic event statistics and the systematics of the selection. The uncertainty of the leptonic pole forward–backward asymmetry is statistics-limited. The predicted value has a quite strong dependence on both m_H and m_t but also on the residual uncertainty of the electromagnetic coupling constant $\alpha(m_Z)$.

In general the individual measurements agree well. Only Γ_Z shows a larger scatter with $\chi^2/\mathrm{d.o.f.} = 7.1/3$, which has a probabilty of 7 %.

The results of the nine-parameter fit, where R and A_{FB}^0 are determined for each species, are consistent with the universality of leptonic Z^0 couplings. This is illustrated in Fig. 2.15, which shows the 68 % probability contours in the R_ℓ–$A_{\mathrm{FB}}^{0,\ell}$ plane both for the combined five-parameter fit and individually for each lepton species.

2.3 Polarization Asymmetries

In the Standard Model parity violation in the process $e^+e^- \rightarrow Z^0 \rightarrow f\overline{f}$ is responsible for the relatively small leptonic charge asymmetry at the Z^0 peak. Much stronger are the effects of parity violation on the polarization of the fermions.

Experimentally, there are two ways to exploit the polarization for a study of parity violation in Z^0 couplings. At LEP, with unpolarized e^+e^- beams, the polarization of the final-state fermions must be measured, which can be done precisely only for tau pairs.[12] The SLC, on the other hand, can be operated with a polarized electron beam, which offers a very direct way to measure the parity violation in the Z^0 couplings.

The difference between the τ production rates at LEP with right (σ_R) and left (σ_L) helicity defines the average tau polarization, $\mathcal{P}_\tau$, which is directly related to the couplings parameter $\mathcal{A}_\tau$ (1.159),

$$\mathcal{P}_\tau = \frac{\sigma_R - \sigma_L}{\sigma_R + \sigma_L} \approx -\mathcal{A}_\tau \, , \tag{2.23}$$

neglecting small effects due to QED and higher-order corrections. Furthermore, the asymmetry $\mathcal{A}_e$ in the e–Z^0 couplings causes a net polarization of the Z^0 even for unpolarized beams, which manifests itself as a dependence of the final-state polarization on the scattering angle θ_{τ^-} between the e^- beam and the final-state τ^- lepton. Similarly to the *charge* forward–backward asymmetry this can be expressed as *polarization* forward–backward asymmetry

$$A_{\mathrm{FB}}^{\mathrm{pol}} = \frac{(\sigma_R^F - \sigma_L^F) - (\sigma_R^B - \sigma_L^B)}{\sigma_{\mathrm{tot}}} \approx -\frac{3}{4}\mathcal{A}_e \, , \tag{2.24}$$

where F and B denote the forward ($\theta_{\tau^-} < 90°$) and backward ($\theta_{\tau^-} > 90°$) directions, respectively.

At the SLC the polarized electron beam causes a *left–right* cross-section asymmetry which is directly related to the asymmetry $\mathcal{A}_e$ of the e–Z^0 couplings

$$A_{\mathrm{LR}} = \frac{1}{P_e}\frac{\sigma_l - \sigma_r}{\sigma_l + \sigma_r} \approx \mathcal{A}_e \, , \tag{2.25}$$

where P_e is the e^- beam polarization, and σ_l and σ_r are the cross-sections measured with a left-handed and a right-handed electron beam polarization, respectively.

Complementary information on the final-state couplings $\mathcal{A}_f$ for a fermion flavour f can be obtained at the SLC from the *left–right* forward–backward asymmetry

$$A_{\mathrm{FB}}^{\mathrm{LR}} = \frac{1}{P_e}\frac{(\sigma_l^F - \sigma_r^F) - (\sigma_l^B - \sigma_r^B)}{\sigma_{\mathrm{tot}}} \approx \frac{3}{4}\mathcal{A}_f \, . \tag{2.26}$$

2.3.1 Tau Polarization

The V–A structure of the charged-current decay of the tau lepton causes a direct relation between its polarization and the angular distribution of the decay products in the τ rest frame.

---

[12] Quark polarization can be measured with prompt heavy baryons. The precision of the measurement [162,163] is limited and it is of interest mainly as a test of heavy-quark effective theory.

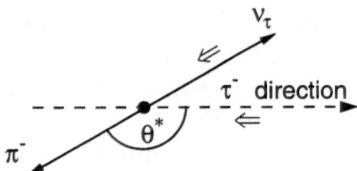

Fig. 2.18. Helicity in the $\tau^- \to \pi^- \nu_\tau$ decay

The effect is most evident for the two body-decay of the τ to a spin-0 meson (π, K) and a τ neutrino, as illustrated in Fig. 2.18. In this decay only the ν_τ can balance the spin angular momentum of the τ. Therefore, in the case of a left-handed τ^-, the ν_τ preferentially goes along the τ flight direction, while for a right-handed τ^- this direction is suppressed. The angular distribution of the meson in the tau rest frame is given by

$$\frac{\mathrm{d}N}{\mathrm{d}\cos\theta^*} \propto (1 + \mathcal{P}_\tau \cos\theta^*) . \tag{2.27}$$

The decay angle θ^* in the τ rest frame can be determined from the energy of the meson in the centre-of-mass energy frame:

$$\cos\theta^* \approx \frac{2E_\pi}{E_{\text{beam}}} - 1 , \tag{2.28}$$

neglecting small corrections due to the τ and π masses.

For other τ decay modes the relation between the polarization and the experimental observables is less direct.

- Semihadronic two-body decay modes to spin-1 mesons (ρ, a_1) are more complicated since two spin configurations of the meson are possible. However, the subsequent decays of the mesons $(\rho \to 2\pi, a_1 \to 3\pi)$ can be used to extract the meson spin state. This way part of the original τ polarization information is regained.
- Leptonic τ decays $(\tau^- \to e^- \bar{\nu}_e \nu_\tau, \tau^- \to \mu^- \bar{\nu}_\mu \nu_\tau)$ are three-body decays. All particles participating have spin 1/2; therefore several helicity configurations are possible. Furthermore, only one particle, the charged lepton, is observed and hence the polarization signal is strongly diluted.

Table 2.7 shows the sensitivity[13] of the τ decay modes together with the branching fraction and the effective statistical weight. Although the $\tau^- \to \pi^-(K^-)\nu_\tau$ is the most sensitive channel, the large $\tau^- \to \rho^- \nu_\tau$ branching ratio gives this channel the dominant weight.

In general the polarization analysis is performed separately for these five channels. In case of the leptonic and π/K decay modes the measured energy is used directly. From fully simulated Monte Carlo events, the two extreme spectra for $\mathcal{P}_\tau = 1$ and for $\mathcal{P}_\tau = -1$ are constructed. The polarization is

---

[13] The sensitivity S quantifies the obtainable precision $\sigma_{\mathcal{P}_\tau}$ for a given number N of measured τ decays; $S \equiv 1/(\sigma_{\mathcal{P}_\tau} \sqrt{N})$.

Table 2.7. Branching ratio, sensitivity and effective statistical weight for the tau decay modes commonly used in the polarization analysis (OPAL [167]). These are idealized numbers for perfect resolution and efficiency. The actual numbers depend on the analysis method and vary slightly among the experiments

	$\tau \to e\bar{\nu}_e\nu_\tau$ $\tau \to \mu\bar{\nu}_\mu\nu_\tau$	$\tau \to \pi\nu_\tau$ $\tau \to K\nu_\tau$	$\tau \to \rho\nu_\tau$	$\tau \to a_1\nu_\tau$ $a_1^\pm \to \pi^\pm\pi^-\pi+$
Branching ratio	0.35	0.12	0.25	0.09
Ideal sensitivity	0.22	0.58	0.49	0.45
Normalized weight	0.12	0.30	0.44	0.13

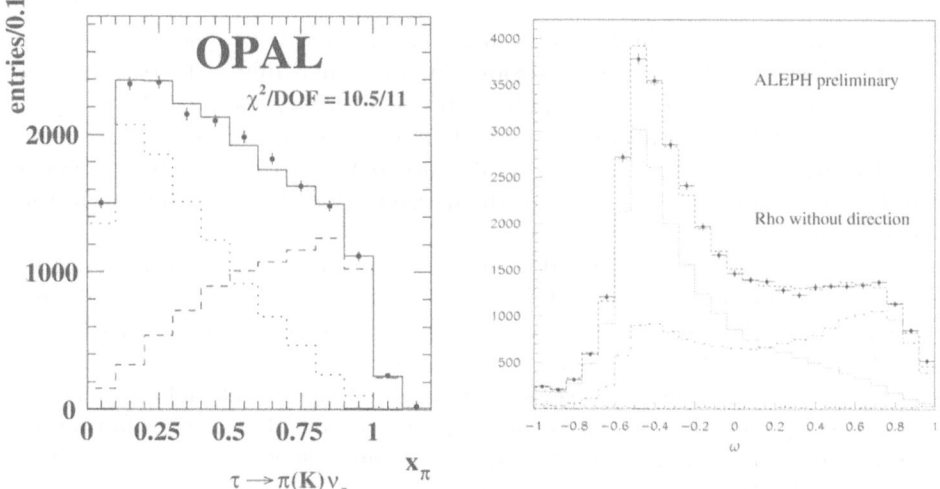

Fig. 2.19. Polarization variable for $\tau^- \to \pi^-(K^-)\bar{\nu}_\tau$ channel by OPAL [167] and for $\tau^- \to \rho^-\bar{\nu}_\tau$ by ALEPH [164]. The histograms show the contributions expected for $\mathcal{P}_\tau = 1$ and $\mathcal{P}_\tau = -1$ and the fitted combination

determined by fitting these to the observed distribution in the data (see Fig. 2.19). Such a fit of Monte Carlo distributions is also performed for the ρ/a_1 channels, but here a fit either multidimensional to the energies and angles of the decay products or to a single kinematic variable which is first constructed from this information is performed.

The τ polarization analysis is rather challenging in terms of detector resolution and performance. In particular, the capabilities of the electromagnetic calorimeter to identify $\pi^0(\to \gamma\gamma)$ are crucial. At high π^0 energies the opening angle of the two photons becomes small, and therefore the overlap of the showers in the electromagnetic calorimeter is increased and they are reconstructed as a single cluster.

Table 2.8. Efficiency and background contamination in the $\tau^- \to \nu_\tau \rho^-$ channel

	ALEPH	DELPHI	L3	OPAL
Efficiency	51 %	41 %	58 %	41 %
Background	8.9 %	15 %	10.5 %	27 %

This makes it difficult to distinguish the decay modes

$$\tau^- \to \nu_\tau \pi^- \,,$$
$$\tau^- \to \nu_\tau \rho^- (\to \pi^- \pi^0) \,,$$
$$\tau^- \to \nu_\tau a_1^- (\to \pi^- \pi^0 \pi^0) \,,$$

which can have very similar signatures: one charged track and closely associated electromagnetic energy. Misidentified channels dilute the polarization signal since the effect of polarization on the variables used for the analysis is rather different in each decay mode. Furthermore, most information in the ρ channel comes from the energy difference between the π^0 and π^-, which depends on the ability to separate the hadronic and photonic showers in the calorimeter.

A fine granularity of the calorimeter and a small amount of material to limit preshowering is important for resolving the two photons of the π^0 and separatin nearby hadronic showers. Longitudinal segmentation (ALEPH and DELPHI) gives additional benefits for identifying hadronic showers. OPAL's design, with relatively large lead glass blocks and heavy material in front, has disadvantages for this measurement, despite its excellent energy resolution. In addition, the low magnetic field reduces the spatial separation of the showers induced by charged and neutral particles.

The performance of the LEP detectors is reflected in Table 2.8, which shows the efficiency and purity for the $\tau^- \to \nu_\tau \rho^-$ channel.

The systematic errors of the measurement arise mainly from uncertainties related to the calorimeter, such as energy resolution and scale and the shower simulation. In general, they can be controlled at a level below the statistical uncertainty.

Figure 2.20 shows the measured polarization as function of the polar angle. The results from all LEP experiments [164–167] for $\mathcal{A}_\tau$, determined from the average τ polarization, and $\mathcal{A}_e$, from the polarization asymmetry, are summarized in Fig. 2.21. $\mathcal{A}_\tau$ and $\mathcal{A}_e$ agree well within the errors. Assuming lepton universality, the combination gives

$$\mathcal{A}_\ell = 0.1450 \pm 0.0033 \,. \tag{2.29}$$

2.3.2 Polarization at the SLC

At the SLC electrons are produced by illuminating a strained gallium arsenide crystal with circularly polarized laser light. Depending upon the sign

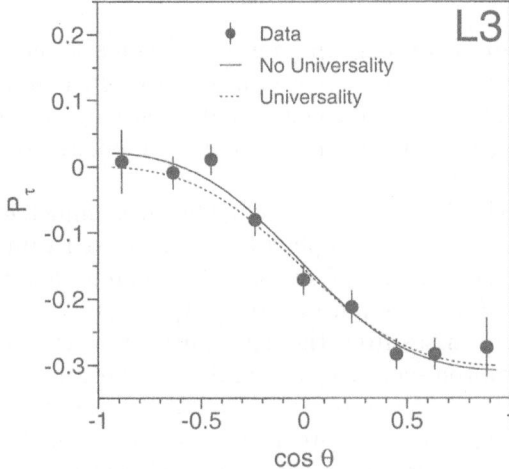

Fig. 2.20. Measured polarization versus $\cos\theta$ (L3 [166]). The *dashed line* shows the result of a fit assuming lepton universality ($\mathcal{A}_\tau = \mathcal{A}_e$), and the *solid line* the fit without this constraint

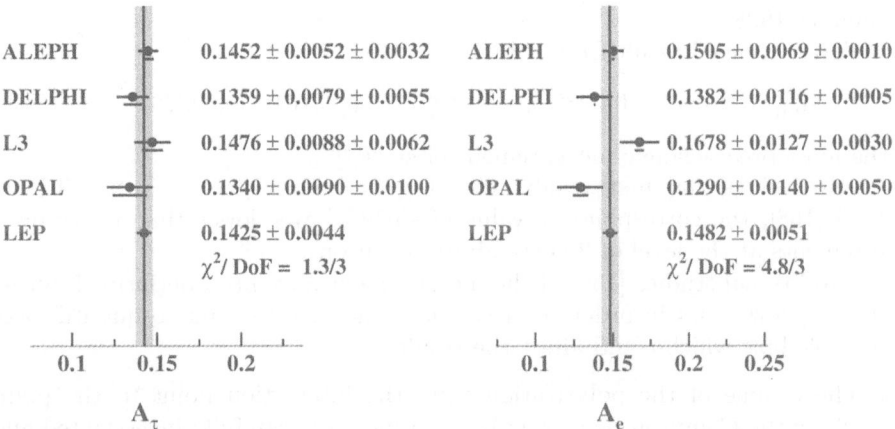

Fig. 2.21. LEP measurements of $\mathcal{A}_\tau$ and $\mathcal{A}_e$ from the average τ polarization $\mathcal{P}_\tau$ and the polarization asymmetry $A_{\mathrm{FB}}^{\mathrm{pol}}$

of the laser polarization predominantly right- or left-handed polarized electrons are emitted and then accelerated. Typically about 88 % of the electrons in the beam have the selected helicity, which corresponds to a polarization ($P_e = (N_{\mathrm{corr}}^{e^-} - N_{\mathrm{wrong}}^{e^-})/N_{\mathrm{total}}^{e^-}$) of the electron beam of 76 %. The sign is switched with each SLC pulse, so that alternately a mostly left- or right-handed electron beam collides with an unpolarized positron beam.

Left–Right Cross-Section Asymmetry

The measurement of A_{LR} is a simple counting experiment based on a selection of hadronic Z^0 decays. The numbers of events produced with left-handed (N_{L}) and right-handed (N_{R}) polarization are counted and essentially A_{LR} is directly given by the asymmetry $(N_{\mathrm{L}} - N_{\mathrm{R}})/(N_{\mathrm{L}} + N_{\mathrm{R}})$ and the polarization P_e according to (2.25).

A few small corrections due to background and residual beam asymmetries (luminosity, energy, polarization) need to be applied to the measured value, and the contributions from photon exchange, γZ interference and initial-state radiation must be subtracted to obtain the pure Z^0 term $A_{\mathrm{LR}}^0 \equiv \mathcal{A}_e$.

The dominant systematic error arises from the determination of the polarization. This is done using Compton scattering of circularly polarized laser light by the electron beam after the collision point; the method is similar to the one used at LEP for the beam energy measurement by resonant depolarization (Sect. 2.2.1). The Compton scattering allows one to closely monitor the average polarization every three minutes with a statistical accuracy of 1 %. The overall systematic error of the polarization is below 1 %, which contributes substantially less to the A_{LR}^0 uncertainty than the statistical error from about 350 000 Z^0 events recorded in polarized running from 1992 to summer 1998.

The averaged result from this data is [169]

$$A_{\mathrm{LR}}^0 = 0.1511 \pm 0.0022 \quad \text{or} \quad \sin^2\theta_{\mathrm{eff}}^{\mathrm{lept}} = 0.23101 \pm 0.00028 , \qquad (2.30)$$

the most precise single measurement of $\sin^2\theta_{\mathrm{eff}}^{\mathrm{lept}}$.

Ever since the first results for A_{LR} had been presented by SLD in 1993 [168], the corresponding value of $\sin^2\theta_{\mathrm{eff}}^{\mathrm{lept}}$ was lower than other measurements at the level of 2–3 standard deviations.

Additional studies [169] of the measurement have been performed during the last few years in order to cross-check the analysis and to quantify any possible bias which could affect the result.

- The change of the polarization from the interaction point to the point where the Compton scattering is performed was carefully investigated and quantified. Such a change can arise owing to different contributions from off-momentum particles to the collision and the Compton scattering.
- The measurement of the Compton electrons was verified with several independent detectors in addition to the original set-up.
- A dedicated experiment demonstrated the negligible level of polarization in the positron beam.
- The beam energy at the SLC is measured with a magnetic spectrometer. In a short scan of the Z^0 resonance m_Z was determined and used to cross-check the energy measurement.

None of these studies revealed unaccounted effects beyond the quoted systematic uncertainties.

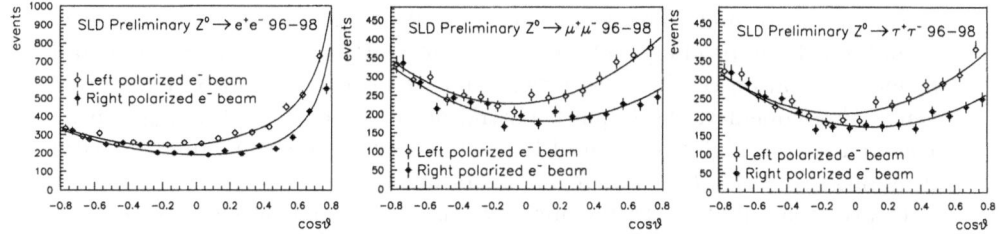

Fig. 2.22. Angular distribution of leptonic events for the two electron beam polarizations (SLD [169])

Left–Right Forward–Backward Asymmetries

The left–right forward–backward asymmetry is measured separately for the three leptonic final states (b and c quark asymmetries are discussed in Sect. 2.5.1). Leptonic final states are selected in the range $|\cos\theta| < 0.8$; the three species are distinguished using similar criteria on multiplicity, charged momentum and total calorimeter energy, as described in Sect. 2.2.3 for the LEP experiments.

Figure 2.22 shows the angular distribution of the three leptons separately for the two polarization states of the electron beam. From a maximum-likelihood fit to these curves both A_{LR} ($\propto \mathcal{A}_e$) and $A_{\mathrm{FB}}^{\mathrm{LR}}$ ($\propto \mathcal{A}_\ell$) were determined. The results are

$$\mathcal{A}_e = 0.1558 \pm 0.0064 \,,$$
$$\mathcal{A}_\mu = 0.137 \pm 0.016 \,,$$
$$\mathcal{A}_\tau = 0.142 \pm 0.016 \,.$$

The uncertainty is completely dominated by statistics. Assuming lepton universality and combining the result with the A_{LR} measurement (2.30) yields the SLD average

$$\mathcal{A}_\ell = 0.1512 \pm 0.0020 \,. \tag{2.31}$$

2.4 Heavy Quarks

The separation of the five quark flavours produced in hadronic Z^0 decays for a precise measurement of the effective couplings is far from straightforward when compared with the charged leptons discussed in Sect. 2.2.3. The hadronization phase largely disguises the original decay flavour and prevents a discrimination based on simple criteria as is possible for the charged-lepton selections.

Instead, more subtle information must be used. Z^0 decays to heavy quarks give experimental handles through the relatively long lifetimes and large masses of the primary heavy mesons as well as the characteristic properties of their decay products. These tagging quantities allow a rather pure and efficient separation of $Z^0 \to b\bar{b}$ events and to some extent also of $Z^0 \to c\bar{c}$ events. Consequently, the b quark partial decay ratio $R_b \equiv \Gamma_b/\Gamma_{\text{had}}$ can be measured with a high precision of below 0.5 %, while for the c quark partial decay ratio $R_c \equiv \Gamma_c/\Gamma_{\text{had}}$ about 3 % is obtained.

In principle, light-quark flavours can also be tagged by identifying their prompt decay products (pions, kaons, protons, Λ^0) [170]. However, the accuracy at the level of 10–20 % obtained for the partial widths by these measurements provides only weak constraints for tests of the Standard Model.

Fortunately, the partial width which can be measured experimentally with the best precision, R_b, is also the most interesting from the theoretical point of view. Since the b quark is the isospin partner of the top, unique contributions of diagrams involving the t quark occur at the $Z^0 \to b\bar{b}$ vertex. In the Standard Model the quantity R_b depends therefore essentially on m_t only, since electroweak corrections of the propagator cancel in the ratio of partial Z^0 decay widths. Moreover, supersymmetric models predict observable changes of R_b for a certain range of model parameters [127]. Additional theoretical interest was triggered by a three-sigma deviation of the measured R_b from the Standard Model prediction in 1995. After the inclusion of additional data and the use of more refined analysis techniques, however, the discrepancy disappeared.

In the next sections, first an overview of the b and c quark tagging methods is presented, followed by a brief description of the R_b and R_c measurements and the heavy-quark forward–backward asymmetries. More details can be found in reviews [173,174].

2.4.1 Heavy-Quark Tagging

Several ways exist to identify decays of the Z^0 to heavy quarks. Already at PEP [171] and PETRA [172] electrons and muons were used to tag heavy-flavour events. High-momentum leptons in hadronic events originate mainly from semi leptonic decays of c and b hadrons. The momentum p and the transverse momentum p_T with respect to the accompanying jet serve to further discriminate the two flavours.

A further possibility is the full or partial reconstruction of a heavy hadron from its decay products. This method is of particular importance for $c\bar{c}$ tagging. Some use is also made of the information about the event shape; the large mass of b hadrons results in a smaller boost and therefore broader jets compared with light-quark events.

However, by far the best tool for tagging heavy-flavour events at LEP and SLC relies on the long lifetime of heavy hadrons. The introduction of semiconductor-based microvertex detectors caused a quantum leap in the lifetime measurements and led to tagging methods with a novel level of purity and efficiency.

Table 2.9. Lifetimes [104], typical impact parameters and decay lengths for b and c hadrons. At energies close to m_Z the typical boost $(\beta\gamma)$ is 7 for b and 12 for c hadrons

	Lifetime τ (ps)	Impact parameter $c\tau$ (mm)	Decay length $\beta\gamma c\tau$ (mm)
B^+	1.54 ± 0.11	0.46	3.2
B^0	1.50 ± 0.11	0.45	3.1
B_s^0	1.34 ± 0.30	0.40	2.8
D^+	1.057 ± 0.015	0.32	3.8
D^0	0.415 ± 0.004	0.12	1.5
D_s^+	0.467 ± 0.017	0.14	1.7

In $Z^0 \rightarrow q\bar{q}$ decays the quark and antiquark are boosted in opposite directions and the subsequent hadronization is largely independent. Therefore most tagging algorithms are not applied on the level of events but independently to the two hemispheres of each event, where the hemispheres are in general defined by a plane orthogonal to the thrust direction.[14]

Lifetime Tagging

The typical vertex resolution of present microvertex detectors is about 300 μm or about a factor 10 smaller than the average b and c hadron decay lengths, which are given in Table 2.9 together with the lifetimes. The lifetime of b hadrons is substantially larger owing to the Cabibbo–Kobayashi–Maskawa suppression of $b \rightarrow c$ transitions. However, since c hadrons have a much lower mass they are boosted more, which leads to rather similar decay lengths of b and c hadrons. The stronger boost also causes on average a narrower cone for the decay products of c hadrons, which reduces the resolution of the measured decay vertex, as illustrated in Fig. 2.23.

The first ingredient of lifetime tagging is a precise reconstruction of the Z^0 decay point or the primary vertex. At LEP the beam spot size is about 120 μm in the horizontal plane, 10 μm in the vertical plane and about 1 cm along the beam line. The nominal position is determined experimentally for short periods by averaging over a few hundred events to follow closely variations with time. The primary vertex of the event is determined by an iterative procedure. First all tracks in an event are fitted to a common vertex using the nominal beam spot as a constraint. Then the track which has the largest

---

[14] The thrust axis is a measure of the event direction and defined as the vector $\boldsymbol{n}$ which maximizes the quantity $T = \sum_i |\boldsymbol{p}_i \boldsymbol{n}| / \sum_i |\boldsymbol{p}_i|$. $\boldsymbol{p}_i$ is the momentum vector of particle i.

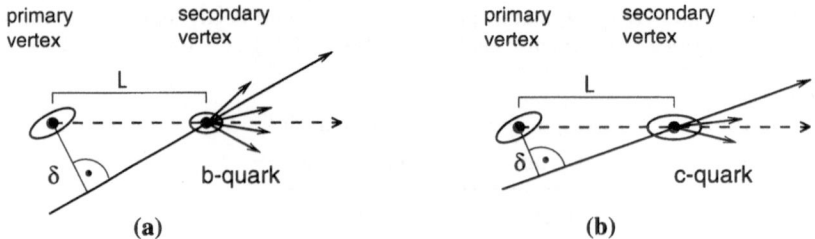

Fig. 2.23. Definition of impact parameter δ and decay length L. b hadons (**a**) are boosted less than c hadrons (**b**). Therefore tracks from b hadrons have a larger impact parameter and the secondary vertex can be determined more precisely

deviation[15] is dropped and the fit is repeated. This procedure continues until no track deviates by more than a certain threshold.

A primary-vertex resolution of about 60 μm in the horizontal plane and about 80 μm along the beam can be reached; vertically the resolution is given by the beam spot constraint of 10 μm.

At the SLC the transverse beam spot region is smaller by orders of magnitude, with a size of (1.5×0.8) μm^2. Therefore the transverse beam spot constraint is sufficient for the primary-vertex determination in the SLD measurements; only the coordinate along the beam is reconstructed, using a similar procedure to the LEP experiments.

Two different approaches are taken by the experiments to exploit the lifetime information for c and b tagging.

Impact Parameter Significance. The impact parameter δ is the distance of closest approach in space of the track and the primary event vertex, as sketched in Fig. 2.23. It is assigned a positive sign if the point of closest approach to the associated jet lies in front of the primary vertex, otherwise, if it is behind, the sign is negative. In order to account implicitly for the uncertainties in the primary vertex position and the track reconstruction it is advantageous to use the impact parameter significance $\mathcal{S}$, the ratio of the impact parameter to its error, δ/σ_δ.

For tracks which originate from particles generated in the hadronization the impact parameter fluctuates randomly around zero, while for a track created in the decay of long-lived particles $\mathcal{S}$ tends to have a positive value. The size of $\mathcal{S}$ is proportional not only to the decay length but also to the mass of the decaying hadron since a higher mass leads to larger decay angles.

Figure 2.24 shows the impact parameter significance as measured by the ALEPH collaboration. The negative part of the $\mathcal{S}$ distribution gives a direct measure of the resolution function, which is used to calculate a probability $\mathcal{P}_\mathrm{T}$ for a given track with a positive $\mathcal{S}$ to be consistent with the primary

---

[15] The difference of the point of closest approach of the track from the fitted common vertex normalized by its uncertainty.

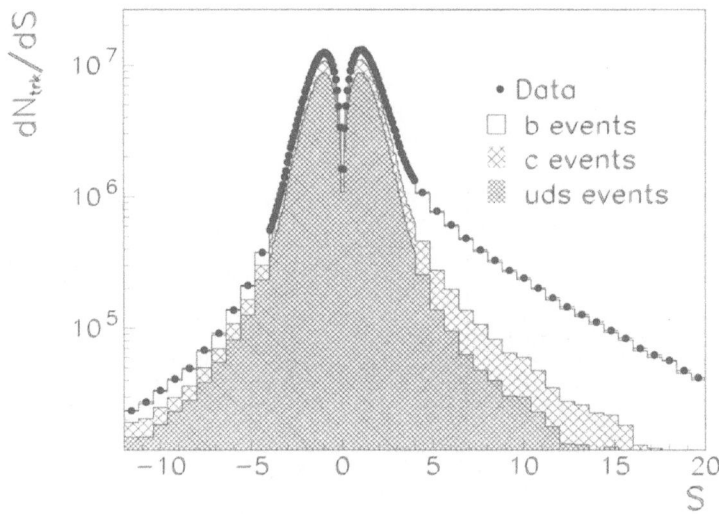

Fig. 2.24. Impact parameter significance S for tracks in data and in uds, c and b Monte Carlo events (ALEPH [175])

vertex. To obtain a variable for the hemisphere tag all $\mathcal{P}_T$ values of tracks with positive S are combined to form a hemisphere probability $\mathcal{P}_H$.

Three of the LEP experiments, ALEPH [175], DELPHI [176] and L3 [177], use such a hemisphere probability as the basic ingredient of the R_b measurement.

Decay Length Significance. The decay length L is the distance from the primary vertex to a secondary vertex (Fig. 2.23). This secondary vertex is reconstructed using a similar iterative algorithm to that for the primary vertex; however, only tracks within one hemisphere or reconstructed jet are used.

Depending on whether the secondary vertex is in front or behind the primary vertex, the decay length is given a positive or negative sign. Similarly to the impact parameter significance, it is more useful to use the decay length significance L/σ_L as a tag variable, since the resolution of the decay length improves with the mass of the decaying hadron.

OPAL [178] and SLD [179] use L/σ_L as basic tagging variable for R_b. Figure 2.25 shows the distribution measured by OPAL. The negative part of the distribution can be used to control the experimental resolution.

It should be emphasized that for both methods the performance of the tag variable scales with the lifetime of the hadrons rather than their decay length. The lower mass of c hadrons, which leads to a reduced difference in decay length owing to the stronger boost, causes a worse secondary-vertex resolution or decreases the impact parameters owing to the smaller decay angles.

The separation of b and c quarks can be further improved by making explicit use of the larger mass of b hadrons. SLD and DELPHI utilize directly

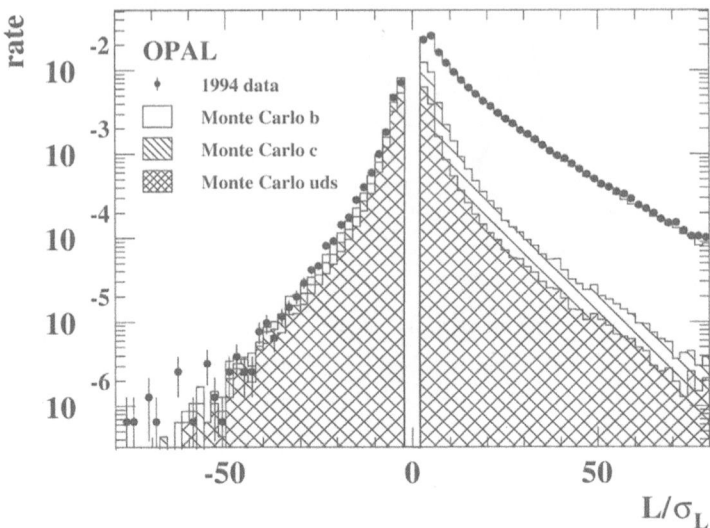

Fig. 2.25. Decay length significance L/σ_L for data and uds, c, and b Monte Carlo events (OPAL [178]).

Table 2.10. Efficiency and background for lifetime tagging of b hemispheres for the LEP experiments and SLD

	ALEPH	DELPHI	L3	OPAL	SLD
Efficiency	19.2 %	29.6 %	25.5 %	23.7 %	50.3 %
Background	1.5 %	1.5 %	16.0 %	2.0 %	2.2 %

the invariant mass constructed from tracks which are consistent with the secondary vertex. ALEPH and OPAL employ a more complicated method: tracks are combined in order of decreasing impact parameter significance $\mathcal{P}_T$ until their invariant mass exceeds the typical c hadron mass of 1.8 GeV. The $\mathcal{P}_T$ of the last track added is used as an additional tag variable.

In the next step the lifetime and mass tags are combined into a single lifetime–mass variable which defines the b quark hemisphere tag. Table 2.10 lists the efficiencies and backgrounds for b hemisphere tagging for the LEP experiments and SLD. The four LEP experiments have efficiencies around 20–30 %. ALEPH, DELPHI and OPAL all have low backgrounds, around 1.5 %, while L3 allows a substantially higher level of 16 %. SLD also sets the working point at a low background of 1.5 % but reaches a much better efficiency of 50 %.

The SLD experiment benefits from the different conditions at a linear collider. The area of the luminous region is smaller by a factor 1000, which gives, in the first place, a very precise constraint on the primary vertex.

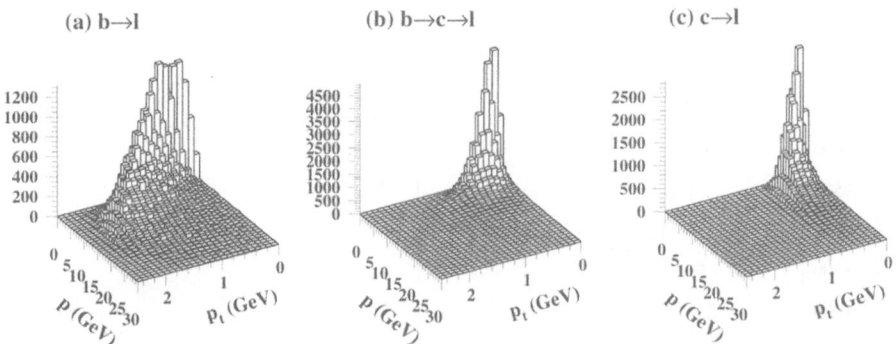

Fig. 2.26. Lepton momentum versus transverse momentum predicted by JETSET Monte Carlo simulation for (**a**) semileptonic b decays, (**b**) cascade decays and (**c**) semileptonic c decays (from [173])

Secondly, it allows a beam pipe with a much smaller radius, which in turn leaves room to install the microvertex detector much closer to the interaction point. The first layer of the SLD detector has a radius of 3 cm, while the LEP detectors start at 6 cm. This significantly improves the lever arm when extrapolating tracks back to the decay point.

Lepton Tagging

In hadronic Z^0 decays electrons and muons originate from the following sources:

- semileptonic b decays, $b \to l$;
- cascade b decays, $b \to c \to l$ and $b \to \tau \to l$;
- semileptonic c decays, $c \to l$;
- instrumental backgrounds: photon conversion, π and K decays to μ and misidentified hadrons.

The different channels can be discriminated by using the momentum p and the transverse momentum relative to the associated jet p_T of the leptons, as illustrated in Fig. 2.26. Semileptonic b decays have the hardest p and p_T spectra, followed by semileptonic c decays and then the cascade b decays. The instrumental background is even softer in both p and p_T.

The efficiency of lepton tags is naturally limited by the branching ratio of approximately 20 % for semileptonic b and c hadron decays. Applying sufficiently stringent criteria for the lepton identification and for p and p_T to reach a purity of 90 % for b quark hemispheres results in an efficiency of typically 6 %.

For asymmetry measurements the semileptonic decays provide a direct way to determine the charge of the quark. However, special care is needed

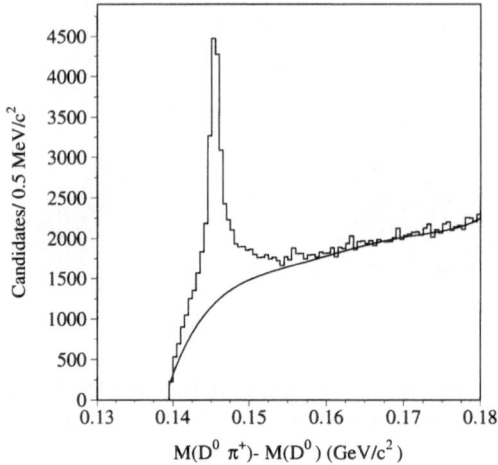

Fig. 2.27. Mass difference $D^* - D^0$. The *solid line* shows the expected background according to the Monte Carlo simulation (ALEPH [180])

for the $b \to c \to l$ cascade decays since the charge of the lepton is inverted with respect to the original quark.

Reconstructed Particles

Reconstructed heavy hadrons provide a rather clean way of tagging $c\bar{c}$ and $b\bar{b}$ events, since heavy hadrons are rarely produced in the hadronization phase. However, an unambiguous identification relies on the full reconstruction of the decay chain, which is feasible only for a few channels. Therefore the efficiency is rather limited owing to the small branching ratios of these identifiable channels and is typically of the order of 1 %.

Particle reconstruction is mainly used for the tagging of $c\bar{c}$ events since for $b\bar{b}$ more powerful tools exist. The discussion here will be restricted to D^* mesons, which are the most important for c tagging.

D^* decays have rather distinctive features which make the identification relatively easy. The decay

$$D^* \to D^0 \pi^+ \tag{2.32}$$

has a branching ratio of about 68 %. The masses of D^* and D^0 differ only by 145.5 MeV, leaving little phase space for the decay products. This mass difference provides a unique signature for D^* identification.

The analysis proceeds by first identifying a D^0 candidate through the decay channels

$$D^0 \to K^- \pi^+ , \ D^0 \to K^- \pi^+ \pi^0 , \ D^0 \to K^- \pi^+ \pi^- \pi^+ ,$$

which account for about 25 % of the D^0 decays. Together with a pion candidate this forms the final D^*. Figure 2.27 shows an example of the reconstructed mass difference between the D^* and D^0. True D^* show up as a sharp peak around 145 MeV, just above the kinematic limit given by the charged-pion mass of 139.6 MeV. It is obvious that good particle identification, namely pion and kaon separation, is very important for this method.

D^* tagging yields a rather pure sample of $b\bar{b}$ and $c\bar{c}$ events. The $b\bar{b}$ component can be further identified using the other tags described above and then statistically subtracted to measure the $c\bar{c}$ contribution.

Three of the LEP experiments, ALEPH, DELPHI and OPAL, based the R_c measurement on D^* reconstruction. At L3 the limited particle identification capabilities of the small central tracking detector make a competitive measurement difficult.

2.4.2 R_b Measurement

The most straightforward way to determine R_b would be "single" tags for each event. The fraction of tagged events is

$$\frac{N_{\text{tag}}}{N_{\text{had}}} = \varepsilon_b R_b + \varepsilon_c R_c + \varepsilon_{uds} R_{uds} , \qquad (2.33)$$

where $\varepsilon_b, \varepsilon_c$ are the efficiencies for tagging $b\bar{b}$, $c\bar{c}$, respectively. ε_{uds} is the combined efficiency for $u\bar{u}$, $d\bar{d}$, $s\bar{s}$. With the Standard Model prediction for R_c, the constraint $R_{uds} = 1 - R_c - R_b$ and the tagging efficiencies from the Monte Carlo simulation, R_b is readily obtained:

$$R_b \approx \frac{N_{\text{tag}}}{\varepsilon_b N_{\text{had}}} , \qquad (2.34)$$

neglecting the c and uds backgrounds, which can be made small. However, the Monte Carlo prediction for ε_b suffers from large systematic uncertainties. Depending on the tag, ε_b is strongly affected by detector parameters, such as resolution, tracking and lepton identification, and by the physics modelling and input, such as the fragmentation function, branching ratios and lifetimes. Altogether this limits the systematic precision to a level of several per cent.

Double Tagging

Given the high event statistics at LEP and the good efficiency of the tags, it is much more advantageous to employ a "double"-tag technique, where the tags are applied on the level of event hemispheres. The number of tagged hemispheres, N_t, is counted and also the number of events with both hemispheres tagged, N_{tt}. These numbers obey

$$N_t = 2N_{\text{had}} \left[\varepsilon_b R_b + \varepsilon_c R_c + \varepsilon_{uds} (1 - R_b - R_c)\right] , \qquad (2.35)$$

$$N_{tt} = N_{\text{had}} \left[(1 + \rho_b)\varepsilon_b^2 R_b + (1 + \rho_c)\varepsilon_c^2 R_c \right. \\ \left. + (1 + \rho_{uds})\varepsilon_{uds}^2 (1 - R_b - R_c)\right] . \qquad (2.36)$$

The correlation parameters ρ_q are introduced to account for possible efficiency correlations between the two hemispheres of an event. These two equations can be solved for R_b and ε_b if the other parameters are taken from Monte Carlo simulation, giving

$$R_b = \frac{N_t^2}{4N_{tt}N_{had}} \, , \quad \varepsilon_b = \frac{2N_{tt}}{N_t} \, , \tag{2.37}$$

neglecting background and correlations.

This method determines ε_b directly from the data, implicitly accounting for all effects of resolution, modelling and simulation. This is at the expense, however, of a much larger statistical uncertainty, given by the number of double-tagged events N_{tt}. At LEP ($N_{had} = 4 \times 10^6$, $\varepsilon_b = 20\%$) the statistical precision is typically 0.5 %.

Systematic Uncertainties

The double-tagging method to determine R_b still relies on the Monte Carlo estimation of the background efficiencies ε_c and ε_{uds} and the correlation coefficients ρ_q.

In particular, the hemisphere correlations have turned out to cause potentially rather large uncertainties. They can arise for a variety of reasons.

- Gluon radiation. Radiation of hard gluons in the perturbative phase reduces the momentum of both b quarks, which results in a smaller decay length of both b jets, leading to a positive correlation between the hemispheres. However, in case of a very hard gluon, both b jets can be reconstructed in the same hemisphere, which gives a negative correlation.
- Geometrical effects. The tracking efficiency, resolution or lepton identification depends on the polar angle θ of the event, owing for example to multiple scattering or detector inhomogeneities; this causes a positive correlation of the hemisphere tagging efficiencies. Furthermore, the rather different resolutions of the primary vertex in the x and y directions introduces positive correlations depending on the azimuthal angle ϕ.
- Common primary vertex. The reconstruction of a common primary vertex for the two hemispheres can cause severe negative correlations. The longer the lifetime of a b hadron in a specific event happens to be, the more likely it becomes that its decay particles are correctly rejected for the primary-vertex determination, and vice versa. This effect pulls the primary vertex, in general, towards the hemisphere with the shorter lifetime. The size of this correlation is difficult to assess precisely; therefore all LEP experiments now avoid this problem by evaluating the primary vertex separately for the two hemispheres, which essentially eliminates the correlation. Although thereby the resolution of the primary vertex is reduced by some 30 % and the tagging performance worsened (typically there is a 10 % relative background increase for the same efficiency), this is more than balanced by the reduced systematic uncertainty.

The dominant background source is $Z^0 \rightarrow c\bar{c}$ events. A large number of contributions enter the estimate and its error: the lifetime of c hadrons, their production rate in Z^0 decays and their decay modes cause uncertainties in the physics modelling. Equally important are errors related to the tracking resolution and efficiency.

The backgrounds from light quarks are at a very low level, typically below 1.5 %. The fact that they contribute at all is mainly due to events where a hard gluon splits into $b\bar{b}$ or $c\bar{c}$, a process which is in principle largely suppressed but theoretically not precisely calculable. Direct measurements of gluon-splitting rates by the LEP experiments [185] were important in constraining this uncertainty.

Multivariate Methods

In order to maximize the use of the information available from the various tag methods, all LEP collaborations have extended the above double-tag method and combined several tags. L3 and OPAL use a rather straightforward extension by combining a lifetime and a lepton tag at the level of hemispheres. This way "mixed" double tags – one hemisphere with a lifetime tag, the other with a lepton tag – increase the number of double-tagged events by some 30 % and improve the efficiency correspondingly.

A more sophisticated approach was pioneered by ALEPH in 1996. Five mutually exclusive tags are applied at hemisphere level. The first three are designed to tag b quark hemispheres:

- tight lifetime–mass tag;
- looser lifetime–mass tag together with neural net event shape information;
- lepton tag.

In addition one tag for c hemispheres (lifetime–mass and event shape) and one tag for light quarks (lifetime and event shape) were developed.

Applying all five tags in a predefined order and mutually exclusively results in five hemispheres rates (N_t^i) and fifteen measurements of mixed-tag events (N_{tt}^{ij}). This leads to a system of 20 equations

$$N_t^i = 2N_{\text{had}} \left[\varepsilon_b^i R_b + \varepsilon_c^i R_c + \varepsilon_{uds}^i \left(1 - R_b - R_c \right) \right] , \tag{2.38}$$

$$N_{tt}^{ij} = N_{\text{had}} \left[(1 + \rho_b^{ij}) \varepsilon_b^i \varepsilon_b^j R_b + (1 + \rho_c^{ij}) \varepsilon_c^i \varepsilon_c^j R_c \right.$$
$$\left. + (1 + \rho_{uds}^{ij}) \varepsilon_{uds}^i \varepsilon_{uds}^j (1 - R_b - R_c) \right] , \tag{2.39}$$

with 62 unknowns: R_b, R_c, 15 tag efficiencies (each tag applied to b, c and uds hemispheres) and 45 correlations.

R_c is fixed at the Standard Model value. The correlations are determined from the Monte Carlo simulations; those which have a significant impact on R_b are subject to more detailed analyses. The background efficiencies for

Table 2.11. R_b error composition in units of 10^{-4}

	ALEPH	DELPHI	L3	OPAL	SLD
Detector resolution	5	1	4	4	10
Correlation	3	3	7	7	4
Physics simulation	7	5	18	7	7
Simulation statistics	5	3	8	1	9
Total systematic	11	6	26	13	14
Data statistics	9	7	15	11	14

the tight lifetime–mass tag are also taken from the Monte–Carlo simulation. Because of the high b purity of this tag, the direct determination would result in a large statistical uncertainty. This leaves R_b and 13 efficiencies, which are determined from a fit to the 20 equations.

The overall complexity of the multivariate approach is significantly higher than that of the rather transparent double tagging technique. But some of the benefits of the multivariate analysis can be qualitatively understood. Compared with the double-tag analysis, the tight lifetime–mass tag is substantially more pure, which reduces systematic uncertainties. The loss in statistics is over-compensated by the gain achieved with the softer b tags. Their background uncertainty can be controlled with good precision by the additional charm and light-flavour tags. Together this leads to a net gain in both statistical and systematic errors.

Compared with the determination of R_b based on the lifetime–mass double-tagging method, the multivariate analysis [175] reduces both the statistical and the systematic uncertainties by about 20 %.

The DELPHI collaboration also applied a multivariate method to determine R_b [176]. The principle of the technique, i.e. several levels of b tags and additional charm and light-flavour tags, is similar to the ALEPH analysis, but differs in the details of the tags and external constraints. DELPHI's quoted statistical and systematic uncertainties exceed the good precision of the ALEPH result by about 25 %, giving the single most precise measurement at the present time.

Table 2.11 gives a breakdown of the error contributions for the R_b measurements of the four LEP collaborations and SLD.

2.4.3 R_c Measurement

The performance for tagging $c\bar{c}$ events at LEP is much worse than for $b\bar{b}$, which leads to a significantly less precise measurement of R_c. In principle, D^* reconstruction allows a rather pure selection of $c\bar{c}$ events (after statistical subtraction of the $b\bar{b}$ component). However, the efficiency is at the level of a

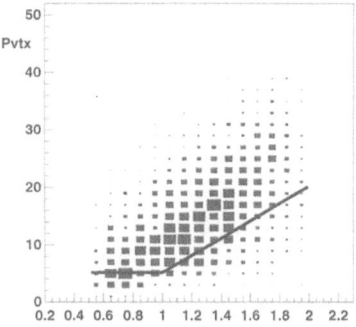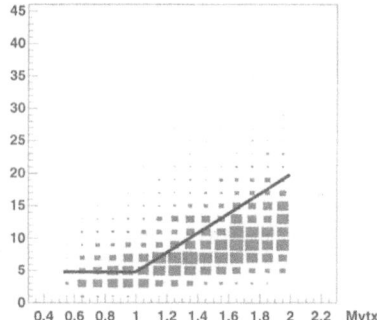

Fig. 2.28. Vertex momentum versus vertex mass for c (*left*) and b (*right*) hemispheres (SLD [183])

few per cent. Therefore double tagging based on this method selects only a few hundred events and gives a correspondingly large statistical error. So far, only ALEPH [180] has used it for R_c, in combination with other methods.

The statistical precision can be improved with a variant of the D^* reconstruction: the exclusive D^* reconstruction in one hemisphere is combined with an inclusive reconstruction in the other, which employs only the characteristic property of the "slow" pion in the D^* decay. Such a mixed tag substantially improves the statistical precision. The efficiencies of the exclusive and inclusive reconstructions must be determined from the Monte Carlo simulation, but since both tags use the $D^* \to D^0\pi^+$ decay the uncertainties associated with $c \to D^*$ production and the D^* decay modes cancel. Further systematic errors arise from hemisphere correlations, the b background subtraction and the uncertainty in the D^0 decay.

A complementary approach is the "charm counting", where all weakly decaying charmed hadrons (D^0, D^+, D_s, Λ_c) are measured by the reconstruction of specific decay modes. This is a single-tag measurement, which directly relies on the charmed hadron decay branching ratios and reconstruction efficiencies. However, it is insensitive to the details of c hadron production rates, since essentially all c quarks end up in one of these states, except for a tiny fraction of charmed baryons.

Charm counting and the inclusive/exclusive D^* double tag are employed by ALEPH [180], DELPHI [181] and OPAL [182] for the measurement of R_c. ALEPH, in addition, has performed a measurement based on lepton tagging.

A very competitive result for R_c has also been obtained by SLD [183]. The excellent performance of the SLD vertex detector allows the construction of a c tag with 15 % efficiency and 69 % purity based on the same lifetime–mass tag as used for the b tagging and additional kinematic cuts on the "vertex" momentum (Fig. 2.28). The good efficiency and purity maks it possible to measure R_c with a double-tag technique. An additional benefit of the method is that the dominant background from b decays can be constrained with data

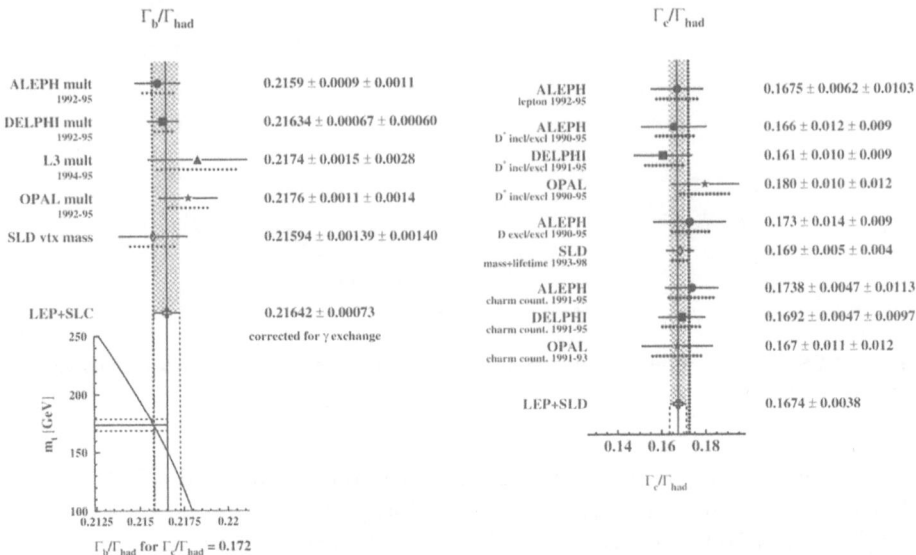

Fig. 2.29. Results of the individual experiments and methods for R_b *(left)* and R_c *(right)*. The error bar on each measurement indicates the total uncertainty; the second error bar below it illustrates the systematic part alone. For R_b, the Standard Model prediction as a function of m_t is also shown

by means of the mixed-tag event rate, where one hemisphere is tagged as b and the other as c. Furthermore, the systematics are largely independent of the LEP R_c measurements.

2.4.4 Results

Figure 2.29 shows the results of the individual experiments for R_b and R_c. The combined averages were determined in a global fit to all data relevant to the heavy-flavour observables [184]. These observables are not only R_b and R_c, but also the pole asymmetries from LEP ($A_{FB}^{0,b}$, $A_{FB}^{0,c}$) and SLD ($\mathcal{A}_b$, $\mathcal{A}_c$) (described in the next section) and the input quantities needed for the measurements, such as production and decay ratios of c and b hadrons and the $B^0 \overline{B^0}$ mixing.

The procedure took into account correlations between the methods, experiments and observables. The LEP collaborations and SLD agreed on a common set of input parameters and their associated uncertainties to facilitate the combination.

The combined results of this fit and the correlations are given in Table 2.12.

Table 2.12. Results and correlations of the heavy-quark combination fit. The heavy quark electroweak observables measured at LEP and SLD are given here. In total, 96 measurements entered the combination and seven input parameters were determined in addition to the six electroweak quantities [184]. The χ^2/d.o.f. of the fit is $56/(96-14)$

		R_b	R_c	$A_{\mathrm{FB}}^{0,b}$	$A_{\mathrm{FB}}^{0,c}$	$\mathcal{A}_b$	$\mathcal{A}_c$
R_b	0.21642 ± 0.00073	1.00	-0.14	-0.03	0.01	-0.03	0.02
R_c	0.1674 ± 0.0038		1.00	0.05	-0.05	0.02	-0.02
$A_{\mathrm{FB}}^{0,b}$	0.0988 ± 0.0020			1.00	0.09	0.02	0.00
$A_{\mathrm{FB}}^{0,c}$	0.0692 ± 0.0037				1.00	-0.01	0.03
$\mathcal{A}_b$	0.911 ± 0.025					1.00	0.15
$\mathcal{A}_c$	0.630 ± 0.026						1.00

2.5 Quark Asymmetries

The measurements of the quark forward–backward asymmetries at LEP and SLD complete the set of precision electroweak measurements at the Z^0 pole.

Owing to the hadronization the determination of the charge of the primary fermion is much less direct than in the case of leptonic final states. Heavy quarks tagged by leptons or particle reconstruction, which yield the charge of the primary quark as a by-product of the tag, are relatively straightforward.

Otherwise, jet charge algorithms can be used, which determine the charge from a weighted sum of the track charges in a jet or hemisphere:

$$Q_{\mathrm{jet}} \equiv \frac{\sum_i Q_i |p_i^{||}|^\kappa}{\sum_i |p_i^{||}|^\kappa} . \tag{2.40}$$

The idea is that the charge of the primary quark manifests itself mostly in the leading hadrons, which in general have a high momentum along the thrust axis ($p_i^{||}$). The exponent κ is chosen to optimize the charge resolution: $\kappa = \infty$ is equivalent to using only the highest-momentum particle, and $\kappa = 0$ corresponds to unweighted charge summing. A principal problem of the jet charge algorithms is the strong dependence on fragmentation modelling. Only when single quark flavours are isolated by additional tags is it possible to control this dependence with the data; otherwise this method introduces large systematic uncertainties.

2.5.1 Heavy-Quark Asymmetries

b and c Asymmetries Using Leptons

In an ideal sample of pure semileptonic $b \to c\,l^-\overline{\nu}$ or $c \to s\,l^+\nu$ decays, the charge sign of the primary quark is directly given by the charge of the tagged lepton and A_{FB} can be determined readily from the angular distribution of the thrust axis:

$$\cos\theta_b = -Q_l \cos\theta_{\mathrm{thrust}}\,, \quad \cos\theta_c = +Q_l \cos\theta_{\mathrm{thrust}}\,. \tag{2.41}$$

In practice several effects dilute the observed asymmetry $A_{\mathrm{FB}}^{\mathrm{obs}}$ in the case of b quarks.

- $B^0\overline{B}^0$ oscillations flip the sign of the lepton charge. $A_{\mathrm{FB}}^{\mathrm{obs}}$ is reduced by $(1-2\chi)$, where χ is the probability for a $b \to \bar{b}$ transition to occur for the selected sample.
- Cascade decays, $b \to c \to l^+$, also cause a sign flip of the observed lepton.
- Background from $c \to l$ decays and misidentified leptons bias $A_{\mathrm{FB}}^{\mathrm{obs}}$.

The lepton-tagged b and c samples are distinguished statistically on the basis p and p_{T} spectra. It is advantageous to determine A_{FB}^b and A_{FB}^c simultaneously by a fit to the three dimensional $(p, p_t, \cos\theta_{\mathrm{thrust}})$ distribution. The resulting measurement is rather precise for A_{FB}^b but much less so for A_{FB}^c, since the contribution of c quarks in the (p, p_{T}) distribution is less pronounced and has a larger background.

b and c Asymmetries Using D^*

c or b events which are tagged by D^* reconstruction also allow a direct charge determination of the primary quark from the charge of the decay products. Since in b events, D^* are produced in cascade decays $b \to c \to D^*$, both c and b quarks lead to the same reconstructed charge, in contrast to lepton tagging. The contributions of b and c events can be separated statistically using the D^* momentum, which tends to be smaller for D^* from b cascade decays. The decay length of the opposite hemisphere or of the D^* measured directly provides additional information to improve the separation.

The precision obtained for A_{FB} is complementary to lepton tagging; rather high for c events, since these are well isolated at large D^* momenta, but low for b events.

b Asymmetry from Jet Charge

In a pure sample of a single quark flavour one can measure the resolution of the jet charge (2.40) from the observed distributions and determine the asymmetry directly.

For N_{F} events with the (negative) b quark in the forward hemisphere and N_{B} events with the b quark in the backward hemisphere the average charge difference between the forward and backward hemispheres $\langle Q_{\mathrm{F}} - Q_{\mathrm{B}}\rangle$ is related to A_{FB}:

$$\langle Q_{\mathrm{F}} - Q_{\mathrm{B}} \rangle = \frac{\big(N_{\mathrm{F}}\langle Q_b \rangle + N_{\mathrm{B}}\langle Q_{\bar{b}} \rangle\big) - \big(N_{\mathrm{B}}\langle Q_b \rangle + N_{\mathrm{F}}\langle Q_{\bar{b}} \rangle\big)}{N_{\mathrm{F}} + N_{\mathrm{B}}}$$

$$= \frac{N_{\mathrm{F}} - N_{\mathrm{B}}}{N_{\mathrm{F}} + N_{\mathrm{B}}} \langle Q_b - Q_{\bar{b}} \rangle = A_{\mathrm{FB}}\, \delta \; . \tag{2.42}$$

The mean charge separation δ is defined as the mean difference between the jet charges of a negative b and a positive $\bar{b}$ ($\delta \equiv \langle Q_b - Q_{\bar{b}} \rangle$).

Assuming no correlation between the forward and backward jet charges and no bias for the average charge, δ is given by

$$\frac{\delta^2}{4} = \langle Q_b \rangle \langle -Q_{\bar{b}} \rangle = -\langle Q_b\, Q_{\bar{b}} \rangle = -\langle Q_{\mathrm{F}}\, Q_{\mathrm{B}} \rangle \; . \tag{2.43}$$

In practice, a slight charge correlation between the two hemispheres arises and the background from other flavours must be taken into account, which is evaluated using simulated events.

But with this technique the observable (A_{FB}^b) and the main correction factor δ_b are determined simultaneously from the data by measuring both the difference $\langle Q_{\mathrm{F}} - Q_{\mathrm{B}} \rangle$ and the product $\langle Q_{\mathrm{F}}\, Q_{\mathrm{B}} \rangle$, similarly to the double-tag method for R_b, which also determines ε_b.

An additional benefit of the method is that it implicitly accounts for effects of $B^0 \overline{B^0}$ oscillations as they are included in the estimate of δ_b from $\langle Q_{\mathrm{F}}\, Q_{\mathrm{B}} \rangle$.

The precision of the measurement is comparable to the measurement using semileptonic decays. The large gain in the number of events due to the good efficiency of lifetime tags is compensated by the much worse resolution of the jet charge measurement compared with the leptonic charge identification.

Heavy-Flavour Asymmetry Results

In order to compare the measured A_{FB} with the theoretical prediction for the "bare" pole asymmetry A_{FB}^0 (1.165), one needs to correct for contributions from γ and γZ interference and account for QED and QCD effects. In particular, a careful treatment of QCD corrections, which are caused mainly by hard gluon radiation, is important as they are large ($\approx 4\,\%$) and depend on the details of the experimental analysis [186].

The LEP combined asymmetries as a function of the centre-of-mass energy are shown in Fig. 2.30. The results of the individual experiments [187–190] corrected to the bare Z^0 pole asymmetries are given in Fig. 2.31. The present averages are

$$A_{\mathrm{FB}}^{0,b} = 0.0988 \pm 0.0020 \; , \tag{2.44}$$

$$A_{\mathrm{FB}}^{0,c} = 0.0692 \pm 0.0037 \; . \tag{2.45}$$

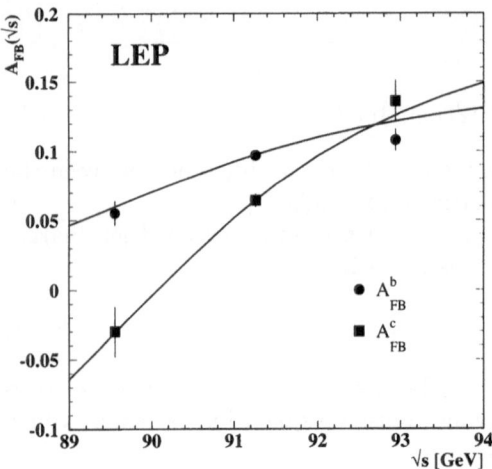

Fig. 2.30. b and c asymmetries measured at LEP versus the centre-of-mass energy. The *solid lines* show the Standard Model predictions

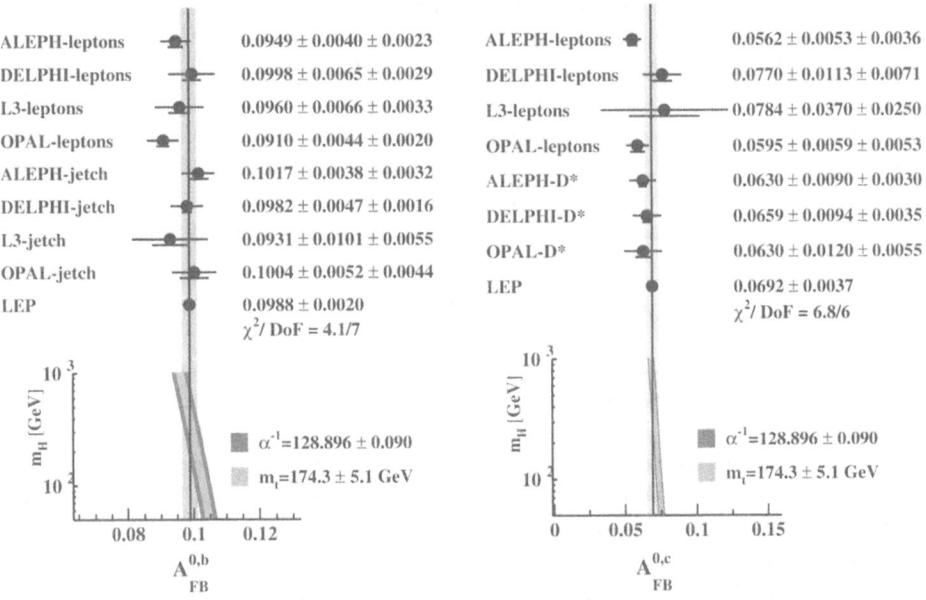

ALEPH-leptons	$0.0949 \pm 0.0040 \pm 0.0023$
DELPHI-leptons	$0.0998 \pm 0.0065 \pm 0.0029$
L3-leptons	$0.0960 \pm 0.0066 \pm 0.0033$
OPAL-leptons	$0.0910 \pm 0.0044 \pm 0.0020$
ALEPH-jetch	$0.1017 \pm 0.0038 \pm 0.0032$
DELPHI-jetch	$0.0982 \pm 0.0047 \pm 0.0016$
L3-jetch	$0.0931 \pm 0.0101 \pm 0.0055$
OPAL-jetch	$0.1004 \pm 0.0052 \pm 0.0044$
LEP	0.0988 ± 0.0020
	$\chi^2 / \mathrm{DoF} = 4.1/7$

$\alpha^{-1} = 128.896 \pm 0.090$
$m_t = 174.3 \pm 5.1 \text{ GeV}$

ALEPH-leptons	$0.0562 \pm 0.0053 \pm 0.0036$
DELPHI-leptons	$0.0770 \pm 0.0113 \pm 0.0071$
L3-leptons	$0.0784 \pm 0.0370 \pm 0.0250$
OPAL-leptons	$0.0595 \pm 0.0059 \pm 0.0053$
ALEPH-D*	$0.0630 \pm 0.0090 \pm 0.0030$
DELPHI-D*	$0.0659 \pm 0.0094 \pm 0.0035$
OPAL-D*	$0.0630 \pm 0.0120 \pm 0.0055$
LEP	0.0692 ± 0.0037
	$\chi^2 / \mathrm{DoF} = 6.8/6$

$\alpha^{-1} = 128.896 \pm 0.090$
$m_t = 174.3 \pm 5.1 \text{ GeV}$

Fig. 2.31. Results of the individual experiments and methods for $A_{\mathrm{FB}}^{0,b}$ *(left)* and $A_{\mathrm{FB}}^{0,c}$ *(right)*. The error bar on each measurement indicates the total uncertainty; the second error bar below it illustrates the systematic part alone. At the bottom of each plot the Standard Model prediction as a function of m_H is shown

Heavy-Flavour Asymmetry at SLD

Owing to the polarized electron beam (Sect. 2.3.2) the heavy-quark asymmetries at the SLD experiment directly measure the final-state coupling parameters $\mathcal{A}_c$ and $\mathcal{A}_b$ with a precision comparable to ($\mathcal{A}_b$) or substantially better ($\mathcal{A}_c$) than the LEP experiments.[16]

SLD uses similar experimental techniques to the LEP experiments, i.e. lepton tags, D^* reconstruction and the jet charge. However, as for the R_c measurement (Sect. 2.4.3), SLD takes advantage of its excellent vertex detector performance and uses a combined lifetime–mass tag and the jet charge for the $\mathcal{A}_c$ determination also. In addition it uses kaons which originate from the secondary vertex and are identified in the Cerenkov ring-imaging detector to determine the charge of the primary quark.

The results [191] are

$$\mathcal{A}_b = 0.911 \pm 0.025 \,, \tag{2.46}$$

$$\mathcal{A}_c = 0.630 \pm 0.026 \,. \tag{2.47}$$

Statistical and systematic errors contribute about equally To the uncertainty of $\mathcal{A}_b$ while for $\mathcal{A}_c$ statistical errors dominate.

2.5.2 Inclusive Quark Asymmetries

The jet charge technique can also be applied to the inclusive sample of hadronic Z^0 decays. The observed forward–backward charge difference is given by the sum of the contributions from the individual quark flavours:

$$\langle Q_\mathrm{F} - Q_\mathrm{B} \rangle = \sum_{q=u,d,c,s,b} \delta_q A_\mathrm{FB}^q \frac{\Gamma_q}{\Gamma_\mathrm{had}} \,. \tag{2.48}$$

However, the asymmetries of the quark flavours A_FB^q all have the same sign, while the mean charge separations $\delta_q \equiv \langle Q_q - Q_{\bar{q}} \rangle$ have opposite signs for u-type and d-type quarks. Therefore the observed inclusive A_FB is strongly reduced since the u and c quark charge asymmetries partially compensate the effect of the d, s and b quark asymmetries.

The individual charge separations δ_q are quite different for each flavour. For heavy flavours (c, b) the separation is in general worse, as the primary hadrons are unstable and the decay products contribute less weight in the jet charge algorithm than the stable high-momentum primary hadrons in light-quark (u, d, s) events.

For heavy quarks the separation (δ_b, δ_c) can be determined from the data by using enriched samples with heavy-flavour tagging. But for light quarks the charge separation relies largely on Monte Carlo simulations, which causes a strong model dependence. In particular, s quark production in the hadronization introduces a large uncertainty.

---

[16] Interpreting the LEP measurements of A_FB^q as $\mathcal{A}_q = 4A_\mathrm{FB}^q/3\mathcal{A}_e$, using the world average for $\mathcal{A}_e$.

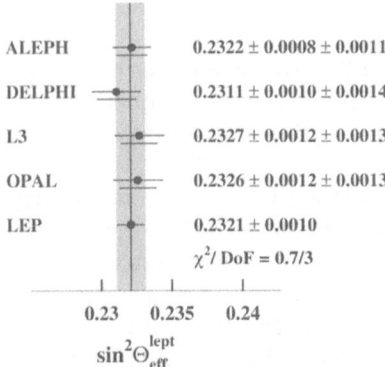

ALEPH $0.2322 \pm 0.0008 \pm 0.0011$

DELPHI $0.2311 \pm 0.0010 \pm 0.0014$

L3 $0.2327 \pm 0.0012 \pm 0.0013$

OPAL $0.2326 \pm 0.0012 \pm 0.0013$

LEP 0.2321 ± 0.0010

             $\chi^2 / \text{DoF} = 0.7/3$

0.23   0.235   0.24

$\sin^2 \Theta_{\text{eff}}^{\text{lept}}$

Fig. 2.32. LEP measurements of inclusive hadronic charge asymmetries

Figure 2.32 shows the inclusive charge asymmetries measured by the LEP experiments [192–195], corrected for electroweak and QCD effects and interpreted directly in terms of $\sin^2 \theta_{\text{eff}}^{\text{lept}}$. The precision is limited by systematic uncertainties in the fragmentation modelling.

2.6 Measurements at LEP2

After the end of the LEP1 programme in 1995 the centre-of-mass energy of the LEP collider was increased step by step. In 1996, for the first time, the threshold for W^+W^- production at 161 GeV was reached. Since then each experiment has recorded about 250 pb^{-1} at high energies, about 75 % of it at 189 GeV in 1998. The centre-of-mass energies and integrated luminosities are summarized in Table 2.1 (Sect. 2.1).

The primary goals of the LEP2 programme are the measurement of the W boson mass and properties, as well as the search for the Higgs boson and other new particles predicted in extensions of the Standard Model.

In addition the measurements of cross-sections and asymmetries for fermion pair production are being continued.

2.6.1 Fermion Pair Production Above the Z

At LEP2 the cross-sections are 2–3 orders of magnitude smaller than at the peak of the Z^0 resonance and the statistical precision is correspondingly reduced. But these measurements still nicely complement the LEP1 results. In the vicinity of the Z^0 resonance the cross-section and asymmetry are rather insensitive to the γZ interference contribution. For the LEP lineshape analysis (Sect. 2.2.4) this term is fixed at the Standard Model prediction. The LEP2 measurements give a good handle to test these assumptions. Similarly,

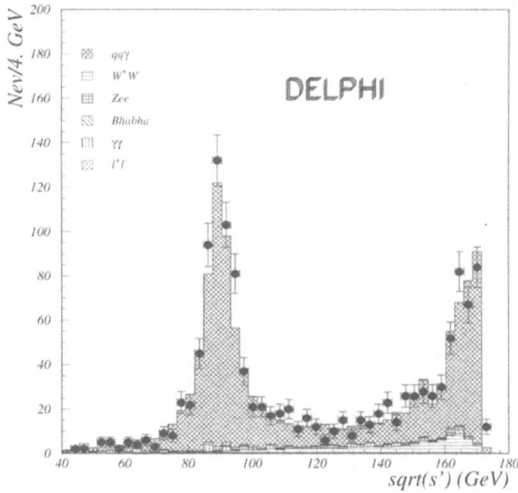

Fig. 2.33. Reconstructed $\sqrt{s'}$ for hadronic events at 172 GeV

new particles mediating interactions between e^+e^- and $f\bar{f}$ could cause effects through interference with the γ and Z^0 diagrams even at energies much lower than their mass scale. The sensitivity to such "new physics" contributions is much higher at LEP2, owing to both the larger centre-of-mass energy and the reduced "background" from Z^0 exchange.

A striking feature of fermion pair production above the Z^0 resonance is the large fraction of events with hard initial-state radiation, where the photons carry away just enough energy and momentum to reduce the effective annihilation energy ($\sqrt{s'}$) to about m_Z.

For $e^+e^- \rightarrow q\bar{q}$ only about 25 % of the events are produced with $\sqrt{s'}$ close to the full centre-of-mass energy. It is important to separate these since the radiative events dilute the sensitivity of the measurement. Experimentally, this can be achieved by a kinematic reconstruction of $\sqrt{s'}$ based on the particle angles and energies. The reconstructed $\sqrt{s'}$ spectrum shows two characteristic peaks, one at $\sqrt{s'} = m_Z$ and the other at $\sqrt{s'} = \sqrt{s}$ (Fig. 2.33). Typically, the resolution for $\sqrt{s'}$ is about 10 %.

The LEP combined cross-section measurements are shown in Fig. 2.34; they are in good agreement with the Standard Model predictions.

The results of a combined fit to the LEP1 and LEP2 cross-section and asymmetry measurements based on the S-matrix ansatz (Sect. 1.4.2, (1.137)) are given in Table 2.13. Figure 2.35 illustrates the gain in precision obtained for m_Z and the hadronic γZ interference ($j_{\text{tot}}^{\text{had}}$) when adding the LEP2 data.

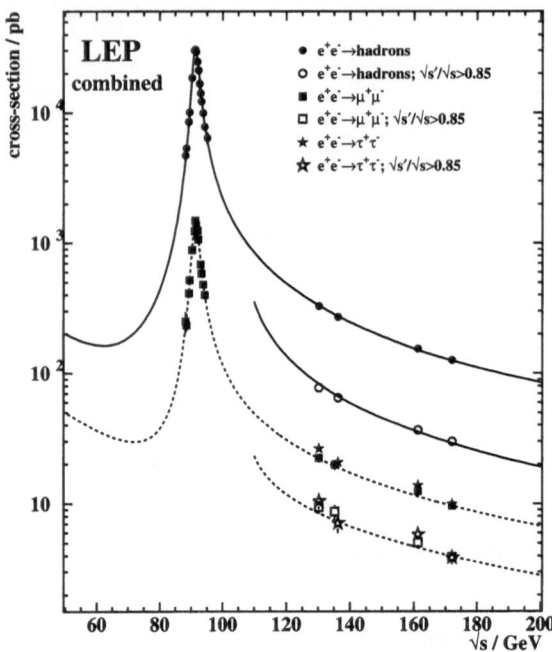

Fig. 2.34. Fermion pair cross-sections measured at LEP

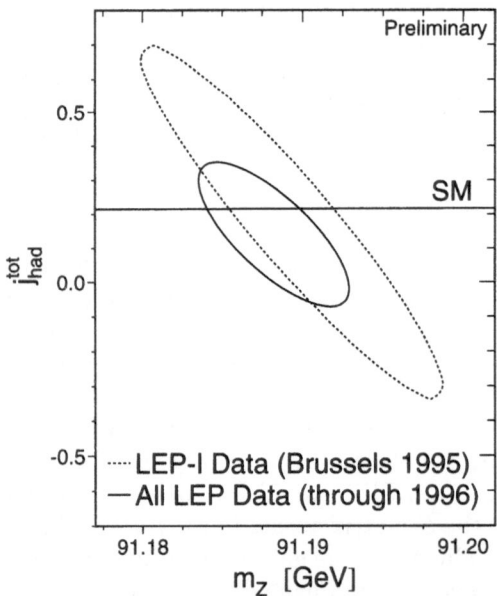

Fig. 2.35. 68 % confidence-level contours for m_Z versus j_{tot}^{had}. The *solid contour* corresponds to all LEP data, while the *dashed contour* represents the status before the start of LEP2 [155]

Table 2.13. Average S-matrix parameters from eight-parameter fits to the data of the four LEP experiments, assuming lepton universality ([155]). m_Z and Γ_Z are corrected to the standard definition of an s-dependent width (1.133). The Standard Model (SM) predictions are listed for $m_Z = 91.1882\,\text{GeV}$, $m_t = 173.8\,\text{GeV}$, $m_H = 300\,\text{GeV}$, $\alpha_s(m_Z^2) = 0.119$ and $1/\alpha(m_Z^2) = 128.878$

	Average value	SM prediction
m_Z (GeV)	91.1882 ± 0.0031	–
Γ_Z (GeV)	2.4945 ± 0.0024	2.4935
$r_{\text{had}}^{\text{tot}}$	2.9637 ± 0.0062	2.9608
r_ℓ^{tot}	0.14245 ± 0.00032	0.14250
$j_{\text{had}}^{\text{tot}}$	0.14 ± 0.14	0.22
j_ℓ^{tot}	0.004 ± 0.012	0.004
r_ℓ^{fb}	0.00292 ± 0.00019	0.00265
j_ℓ^{fb}	0.780 ± 0.013	0.799

2.6.2 Mass of the W Boson

W pairs decay into purely hadronic ($4q$), semileptonic ($2ql\nu$) and purely leptonic ($l\nu l\nu$) final states. All channels have clear experimental signatures and can be selected with efficiencies and purities around 80–90 %. The semileptonic and hadronic channels both have a branching ratio of about 45 % and are therefore most important for the measurement of m_W.

For the first data collected at LEP2 in 1996 the centre-of-mass energy was chosen at 161 GeV, just slightly above the W^+W^- production threshold. Close to this threshold the cross-section for W^+W^- production is rather sensitive to m_W, as illustrated in Fig. 2.36. This way the measured cross-section can be translated into a mass measurement.

At higher centre-of-mass energies the cross-section becomes insensitive to the mass but instead the larger event rate allows a precise mass measurement from the reconstructed kinematic properties of the final states. In the case of the $4q$ channel the mass can be readily obtained from the invariant mass of the jet pairs and similarly for the semileptonic decays, where the neutrino momentum vector can be approximated by the missing-momentum vector. The resolution of the invariant mass can be greatly improved by employing the precisely known centre-of-mass energy at LEP in a kinematic fit of the measured jet or lepton energies and momenta using energy and momentum conservation as constraints. An example of the reconstructed m_W spectrum is shown in Fig. 2.36.

At first one might expect that the $4q$ channel would allow a better measurement since the observed jets from the four quarks provide information for both W bosons, while in the $2ql\nu$ channel the leptonically decaying W is

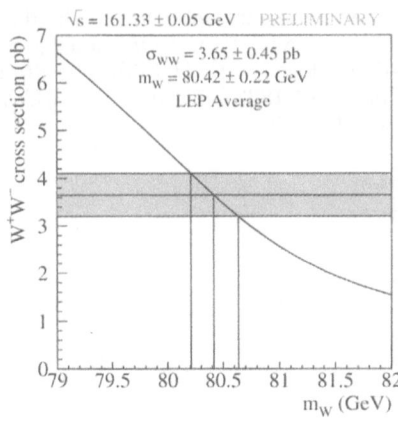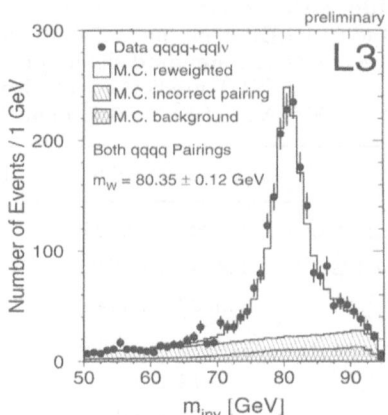

Fig. 2.36. *Left*: cross-section for W^+W^- production at 161.3 GeV versus m_W [196]. *Right*: reconstructed m_W spectrum for the $2ql\nu$ and the $4q$ channel combined (L3 [202])

only partially reconstructed owing to the neutrino. In practice, however, this is not the case:

- energy and momenta are measured much more precisely for electrons and muons than for hadronic jets;
- in four-jet events there are three possible jet pairings for the assignment to the two W bosons, which results in a combinatorial background;
- the background is larger for $4q$ final states ($Z^0/\gamma \to q\bar{q} \to 4$ jets).

Furthermore, final-state interactions between hadrons originating from different W bosons (Bose–Einstein correlation [197] and colour reconnection [198]) potentially bias the mass measurement of the $4q$ final state and lead to larger systematic uncertainties.

Figure 2.37 shows the results of the LEP experiments for the data collected at 189 GeV in 1998. The combined average including all LEP2 data [199–203] is

$$m_W = 80.350 \pm 0.056 \text{ GeV} . \tag{2.49}$$

2.6.3 Search for the Higgs Boson

The Standard Model Higgs boson is expected to be produced at LEP2 predominantly by "Higgs radiation" ($e^+e^- \to Z^{0*} \to Z^0 H^0$, Fig. 2.38). In this mode the Higgs boson should be found up to a natural limit for its mass of

$$m_H \leq \sqrt{s} - m_Z . \tag{2.50}$$

For larger m_H the cross-section falls very rapidly as it requires the Z^0 to be produced off shell. Higgs production via WW or ZZ fusion processes

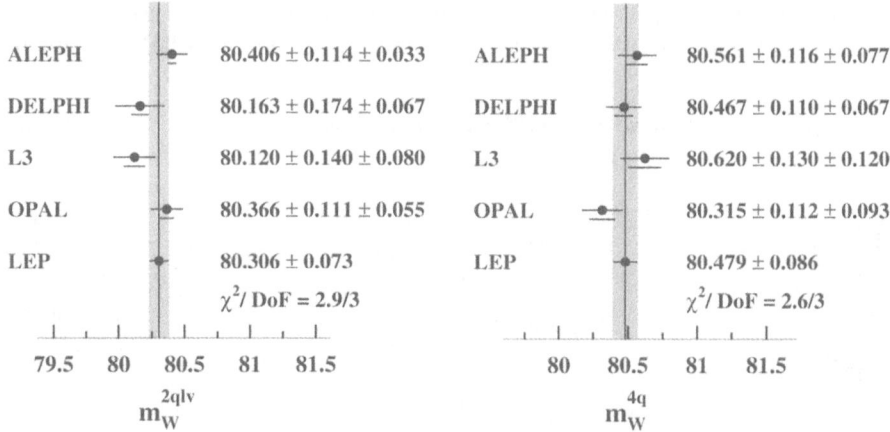

Fig. 2.37. Results for the reconstructed m_W from the LEP experiments for the $2ql\nu$ and the $4q$ channel for the data collected at 189 GeV [200–203]

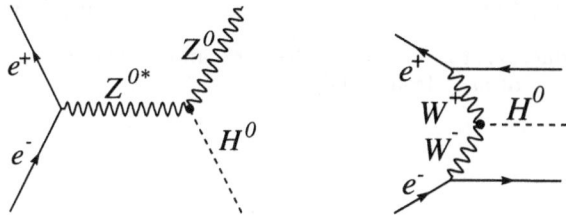

Fig. 2.38. Possible diagrams for Higgs production at LEP2: Higgs radiation (*left*) and WW/ZZ fusion (*right*)

(Fig. 2.38) has a larger cross-section above this threshold, as shown in Fig. 2.39. However, the cross-section is too small to produce a detectable signal for m_H above the limit (2.50) with the luminosity expected at LEP2. The coupling of the Higgs boson to fermions depends strongly on the mass; it is expected to decay predominantly into $b\bar{b}$ (86 %). Therefore b tagging is the crucial ingredient of the analysis, in order to enrich the signal fraction. The three most promising final-state topologies follow from the Z^0 decay branching ratios:

1. four jets: $H^0 \to b\bar{b}$, $Z^0 \to q\bar{q}$;
2. missing energy: $H^0 \to b\bar{b}$, $Z^0 \to \nu\bar{\nu}$;
3. leptonic: $H^0 \to b\bar{b}$, $Z^0 \to e^+e^-$ or $\mu^+\mu^-$.

In general, backgrounds from $e^+e^- \to Z^0/\gamma \to q\bar{q}$, W^+W^-, ZZ and other four-fermion processes contribute to these decay channels. Most of them can be greatly reduced by tight requirements on the b tagging and the final-state kinematics, except for ZZ production, which forms an irreducible background in the mass range accessible to LEP2. The mass distribution of Higgs candidate events observed by the four LEP experiments is shown in Fig. 2.40. The lower limit for m_H derived from the combination of the four LEP experiments is [205–209]

$$m_H = 95.2 \, \text{GeV} \quad \text{at the 95 \% confidence level.} \tag{2.51}$$

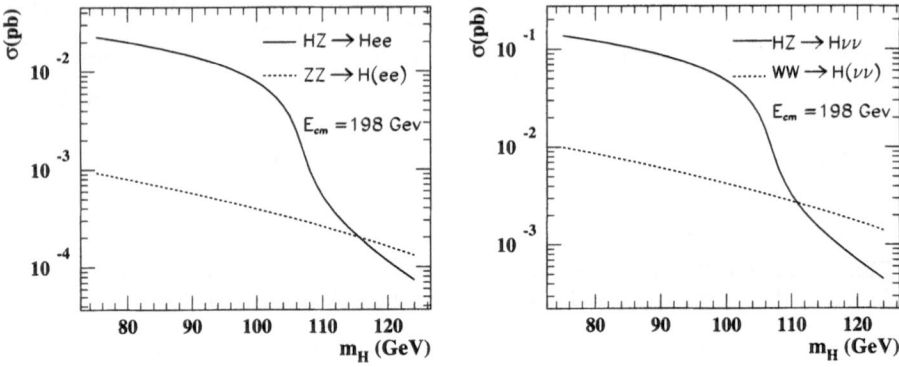

Fig. 2.39. Predicted Higgs production cross-section in the e^+e^- and missing-energy channels at 198 GeV as a function of m_H (from [204])

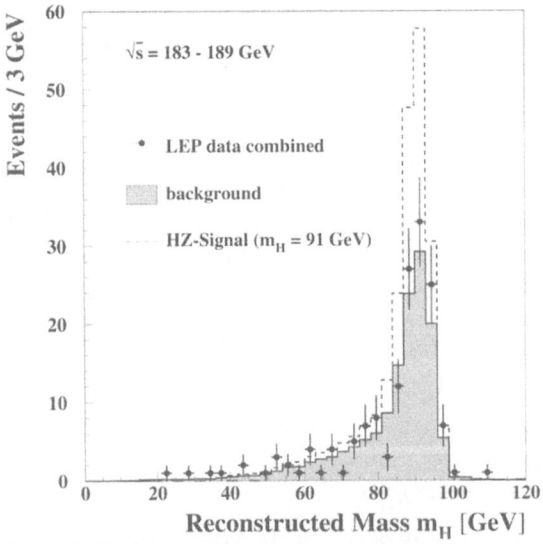

Fig. 2.40. Distribution of reconstructed Higgs candidate masses at LEP. The data (*points*) is compared to the Monte Carlo expectations for the various background processes. Also shown is the expected signal for $m_H = 91$ GeV [209]

2.7 Other Electroweak Measurements

2.7.1 The Mass of the Top Quark

The direct measurement of the mass of the top quark, m_t, is of fundamental importance in the Standard Model. Its large mass – compared to that of the b quark – is responsible for radiative corrections to electroweak observables which in general dominate other effects. Therefore the direct measurement of m_t provides a critical test of the Standard Model and it gives vital constraints to allow the investigation of more subtle effects such as radiative corrections due to m_H.

The first evidence for top production was reported by CDF in 1994 [210] and soon afterwards confirmed by more refined measurements at both CDF and D0. Now, m_t is known with a much better relative precision than for any other quark.

At the TEVATRON, the top is produced predominantly in pairs by the process $q\bar{q} \to t\bar{t} + X$. They decay immediately via $t \to Wb$, $W \to l\nu$ or $q\bar{q}'$. The characteristic signature is therefore a hadronic jet tagged as a b jet accompanied by a pair of jets or a lepton and missing energy due to the neutrino, which are consistent with the decay of a W boson.

The cleanest signal is obtained when the W bosons from both top quarks decay into an electron or muon. However, owing to the W decay branching ratios this mode accounts for only about 5 % of the $t\bar{t}$ events. Moreover, the mass measurement is complicated by the presence of two neutrinos; therefore it contributes only little to the m_t measurement. The mixed channel where one of the W bosons decays into hadron jets is much more frequent (30 %) and can still be selected rather purely. This channel carries most weight (about 75 %) for m_t. The fully hadronic channel has the largest branching fraction (44 %) but suffers from large background and the combinatorial difficulties in the assignment of the six jets to the top and W pairs.

Table 2.14. Top mass results (in GeV, statistical and systematic errors) from D0 and CDF for the individual channels [211]

	Lepton + jets	Di lepton	Hadronic	Combined
D0	$173.3 \pm 5.6 \pm 5.5$	$168.4 \pm 12.3 \pm 3.6$		$172.1 \pm 5.2 \pm 4.9$
CDF	$175.9 \pm 4.8 \pm 4.9$	$167.4 \pm 10.3 \pm 4.8$	$186 \pm 10 \pm 8$	$175.3 \pm 4.1 \pm 5.0$

The results from CDF and D0 are summarized in Table 2.14. The dominating systematic uncertainties are due to the energy scale for the hadronic jets and the physics modelling of top production and of the backgrounds. The combined average is [212]

$$m_t = 174.3 \pm 5.1 \,\text{GeV}\,. \tag{2.52}$$

Common systematic uncertainties arise from the physics modelling and are accounted for in the averaging procedure.

2.7.2 W Mass Measurements at Hadron Colliders

The discovery of the W boson at the UA1 experiment in 1983 [213] also resulted in the first direct measurement of its mass (81 ± 5 GeV). The investigation of the properties of the W continued to be one of the main activities at hadron colliders and a large number of W events has been collected since then, first by the experiments at the CERN Sp$\overline{\text{p}}$S collider and then by D0 and CDF at the TEVATRON.

At hadron colliders much less information is available to reconstruct m_W compared with an e^+e^- collider.

- No direct relation exists between the energy of the W and the centre-of-mass energy.
- Hadronic W decays are difficult to trigger and identify; therefore only leptonic decays are used.
- The longitudinal component of the momentum is not constrained and hence only the transverse neutrino momentum can be estimated.

Therefore, at hadron colliders m_W cannot be directly measured on an event-by-event basis but must be extracted with model-based fits from kinematic variables sensitive to m_W, such as the total or transverse charged-lepton momentum or a transverse mass m_W^{T} constructed from the charged-lepton and neutrino transverse momenta.

The D0 and CDF collaborations each recorded about 200 000 $W \to e\nu$ and $W \to \mu\nu$ events from 1988 to 1996. Figure 2.41 shows the m_W^{T} distribution measured by D0. m_W is determined from a likelihood fit using Monte Carlo spectra generated for different values of m_W.

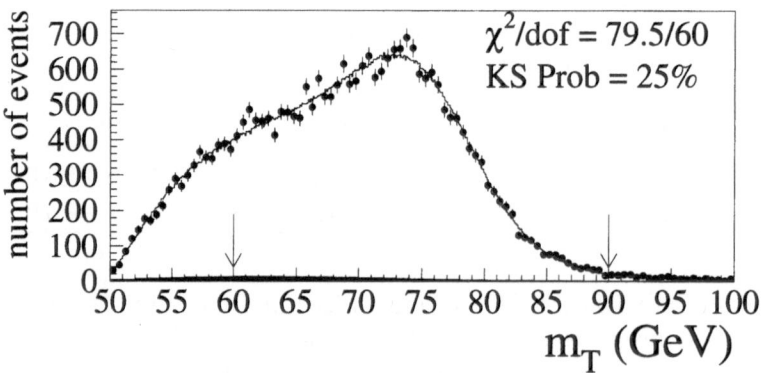

Fig. 2.41. Distribution of transverse W mass m_W^{T} (D0 [214])

A crucial part of the measurement is the energy calibration of the detector. The precisely known masses of the π^0, J/Ψ and Z^0 are very helpful for this purpose. Using these as constraints, the calibration can be performed over a large energy range directly with the collected data, and the residual uncertainty for m_W can be controlled at the level of 30–40 MeV. Uncertainties in the proton structure functions and the modelling of the transverse momentum spectrum of the W bosons produced introduce systematic errors of about 40 MeV, which are largely correlated between the experiments.

The results for m_W from the hadron colliders are [215]

$$m_W \text{ (UA1)} = 80.360 \pm 0.370 \text{ GeV} ,$$
$$m_W \text{ (CDF)} = 80.433 \pm 0.079 \text{ GeV} ,$$
$$m_W \text{ (D0)} = 80.474 \pm 0.093 \text{ GeV} ,$$
$$m_W \text{ (combined)} = 80.448 \pm 0.062 \text{ GeV} . \tag{2.53}$$

2.7.3 Neutrino–Nucleon Scattering

The experiments CHARM and CDHS at CERN, as well as CCFR and its successor NUTEV at Fermilab, measure neutrino–nucleon scattering with both a neutrino and an antineutrino beam. Using the rate of reactions with and without a charged lepton in the final state, one can disentangle charged-current reactions mediated by W bosons and neutral-current reactions mediated by Z^0 bosons. The Paschos–Wolfenstein relation [216]

$$R^- = \frac{\sigma_{NC}^\nu - \sigma_{NC}^{\bar\nu}}{\sigma_{CC}^\nu - \sigma_{CC}^{\bar\nu}} = \frac{1}{2} - \sin^2\theta_W \tag{2.54}$$

relates the measurement to the electroweak mixing angle. This ratio is a rather robust observable as many experimental and theoretical uncertainties are suppressed in a ratio of cross-sections. The differences $\sigma_{NC}^\nu - \sigma_{NC}^{\bar\nu}$ and $\sigma_{CC}^\nu - \sigma_{CC}^{\bar\nu}$ have the additional advantage of minimizing the theoretical uncertainties of ν scattering at sea quarks. Moreover, radiative corrections to the ratio are small; the measured $\sin^2\theta_W$ corresponds closely to the on-shell definition $\sin^2\theta_W = 1 - m_W^2/m_Z^2$.

The most precise measurement was recently performed at NUTEV [72]. Combining it with the earlier CCFR result [71] yields

$$\sin^2\theta_W = 0.2255 \pm 0.0019 \,(\text{stat.}) \pm 0.0010 \,(\text{sys.}) . \tag{2.55}$$

The result has additional slight dependences on m_t and m_H, which are taken into account in the Standard Model fits. Using the LEP1 measurement of m_Z, it can be translated into an indirect measurement of $m_W = 80.25 \pm 0.11$ GeV.

3. Interpretation of the Measurements

In the previous chapter the electroweak precision experiments were discussed. The results are summarized in Table 3.1. This chapter presents interpretations of the measurements in several stages.

First, general predictions of the Standard Model are tested, namely the universality of couplings within one fermion family and the presence of radiative corrections which modify the tree-level expectations of experimental observables.

In the next stage the LEP lineshape results are used to determine the number of light neutrino species, which is, within the Standard Model, equivalent to the number of fermion generations, a fundamental free parameter of the theory.

Finally, free or weakly constrained parameters of the Standard Model are determined by fitting the experimental precision data to the theoretical predictions. In the past such analyses successfully predicted the mass of the top quark, which affects electroweak observables through radiative corrections depending quadratically on m_t. Nowadays, the most important unknown quantity of the Standard Model is the mass of the Higgs boson. In contrast to the top quark, the electroweak radiative corrections depend only logarithmically on m_H. Thus the sensitivity or the predictive power of the measurements is much lower in this case.

Furthermore, consistency tests of the theory can be performed by comparing directly measured observables with an indirect determination in a fit to all other measurements.

3.1 Fermion Couplings

A fundamental prediction of the Standard Model is the universality of the gauge couplings between fermions and bosons. For the effective couplings as definded in (1.122) this is no longer true, since mass- and flavour-dependent radiative corrections occur. But except for b quarks these effects are tiny, well below the experimental precision, so that in practice the effective coupling constants g_V and g_A should be independent of the fermion generation.

Table 3.1. Summary of precision electroweak measurements [155,161]. *Column 2* gives the experimental results and their total uncertainty. *Column 3* lists the systematic part of the error. Since most results are obtained from a weighted average of several measurements with correlated and uncorrelated systematic sources, this is only an approximation meant to illustrate the importance of systematics for each result. The Standard Model predictions in *column 4* and the pulls (difference between measurement and fit in units of the total measurement error) in *column 5* correspond to the parameters determined in the fit to all measurements as discussed in Sect. 3.4

	Result with total error	Systematic error	Standard Model fit	Pull
LEP				
Lineshape and				
lepton asymmetries:				
m_Z (GeV)	91.1871 ± 0.0021	0.0017	91.1869	0.08
Γ_Z (GeV)	2.4944 ± 0.0024	0.0012	2.4957	-0.56
σ_{had}^0 (nb)	41.544 ± 0.037	0.025	41.479	1.749
R_ℓ	20.768 ± 0.024	0.020	20.740	1.156
$A_{\mathrm{FB}}^{0,\ell}$	0.01701 ± 0.00095	0.00060	0.01625	0.80
τ polarization:				
$\mathcal{A}_\tau$	0.1425 ± 0.0044	0.0027	0.1472	-1.07
$\mathcal{A}_e$	0.1483 ± 0.0051	0.0010	0.1472	0.21
$q\bar{q}$ charge asymmetry:				
$\sin^2\theta_{\mathrm{eff}}^{\mathrm{lept}}$ ($\langle Q_{\mathrm{FB}}\rangle$)	0.2321 ± 0.0010	0.0008	0.23150	0.60
SLD				
$\sin^2\theta_{\mathrm{eff}}^{\mathrm{lept}}$ (A_{LR})	0.23099 ± 0.00026	0.00018	0.23150	-1.95
LEP/SLD heavy flavour				
R_b	0.21642 ± 0.00073	0.00057	0.21583	0.81
R_c	0.1674 ± 0.0038	0.0030	0.1722	-1.27
$A_{\mathrm{FB}}^{0,b}$	0.0988 ± 0.0020	0.0010	0.1032	-2.20
$A_{\mathrm{FB}}^{0,c}$	0.0692 ± 0.0037	0.0022	0.0738	-1.23
$\mathcal{A}_b$	0.911 ± 0.025	0.020	0.935	-0.95
$\mathcal{A}_c$	0.630 ± 0.026	0.020	0.668	-1.46
LEP2				
m_W (GeV)	80.350 ± 0.056	0.044	80.385	-0.62
$p\bar{p}$ **and** νN				
m_W (GeV)	80.448 ± 0.062	0.050	80.385	1.02
$\sin^2\theta_W$ (νN)	0.2254 ± 0.0021	0.0010	0.2229	1.13
m_t (GeV)	174.3 ± 5.1	3.9	173.1	0.23

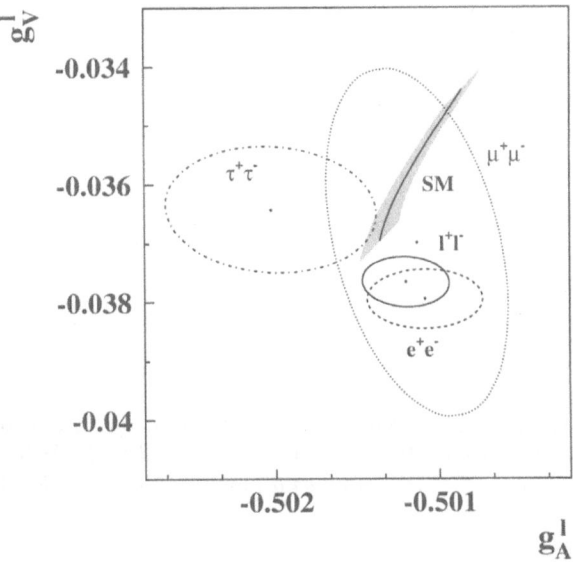

Fig. 3.1. Fit contours (one sigma) of g_V^ℓ versus g_A^ℓ. The *dashed* and *dotted contour lines* result from a fit without assuming lepton universality, while this constraint is imposed for the *solid contour*. The *band* shows the Standard Model prediction for $95 < m_H < 1000$ (*dark line*) and $m_t = 174.3 \pm 5.1$ GeV (*shaded area*)

3.1.1 Lepton Couplings

For the leptonic couplings the experimental measurements allow precise tests of this assumption. The first such test was presented for the LEP line-shape parameters R_ℓ and $A_{\rm FB}^{0,l}$ in Sect. 2.2.6. Combining these measurements with the τ polarization results (Sect. 2.3.1) and the SLD $A_{\rm LR}$ measurement (Sect. 2.3.2) allows the determination of the effective couplings[1] g_V^ℓ and g_A^ℓ, using the relations (1.149), (1.159):

$$\Gamma_\ell \propto (g_V^\ell)^2 + (g_A^\ell)^2 , \quad \mathcal{A}_\ell = \frac{2g_V^\ell g_A^\ell}{(g_V^\ell)^2 + (g_A^\ell)^2} .$$

The one-sigma contours[2] in the (g_V^ℓ, g_A^ℓ) plane determined from the LEP and SLD measurements are shown in Fig. 3.1. For leptons g_A^ℓ and g_V^ℓ are numerically very different; $g_A^\ell \approx 0.5$ and $g_V^\ell \approx 0$. Therefore the partial widths are mainly sensitive to g_A^ℓ, while the asymmetries measure g_V^ℓ.

The electron partial width Γ_e contributes to the cross-sections of all final states. Hence Γ_e or g_A^e has a higher precision than g_A^μ or g_A^τ.

The accuracy for g_V^ℓ is quite different for the three lepton classes, owing to the various asymmetry measurements which contribute. For g_V^μ there is

---

[1] Only the real parts of the effective couplings (1.122) are determined here. The imaginary parts are fixed at the Standard Model predictions.

[2] In two dimensions one sigma corresponds to a 39 % confidence level.

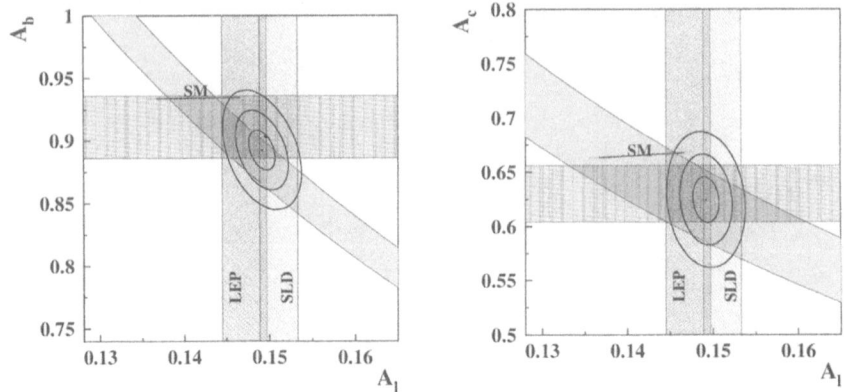

Fig. 3.2. Fit contours (one, two and three sigma) of $\mathcal{A}_b$ (*left*) and $\mathcal{A}_c$ (*right*) versus $\mathcal{A}_\ell$. Also shown are the direct measurements of $\mathcal{A}_b$, $\mathcal{A}_c$ (*horizontal band*), $\mathcal{A}_\ell$ from LEP and SLD (*vertical bands*), $A_{\rm FB}^{0,b}$, $A_{\rm FB}^{0,c}$ (*diagonal band*) and the Standard Model prediction (*horizontal line*)

only the LEP forward–backward asymmetry; τ polarization additionally constrains g_V^τ; and both τ polarization and $A_{\rm LR}$ improve g_V^e, which is therefore determined with the highest precision.

In principle, all the above measurements are symmetric for g_V^ℓ and g_A^ℓ; the results would not change if g_V^ℓ and g_A^ℓ were swapped. But fits to the lineshape and asymmetries without constraining the leptonic γZ interference [159] can resolve this ambiguity. Furthermore, the experiments cannot determine the overall sign of g_V^ℓ and g_A^ℓ. By convention g_A^e is negative; then the signs of all other fermion couplings follow from the measurements.

The results for the individual lepton species are consistent within the uncertainties, supporting lepton universality at the per mille level for g_A^ℓ and at the per cent level for g_V^ℓ. However, both g_V^e and g_V^ℓ are low compared to the Standard Model expectation, about 1.5 sigma away from the predicted value for $m_H = 95$ GeV. This effect is mainly caused by the very precise measurement of $A_{\rm LR}$ which is about two sigma above the Standard Model prediction.

3.1.2 Coupling Parameters

The b and c asymmetries determine at LEP measure the product of the lepton and quark couplings, $A_{\rm FB}^q = \frac{3}{4}\mathcal{A}_e \mathcal{A}_q$. In conjunction with the pure leptonic asymmetries obtained from LEP (assuming lepton universality) and SLD, they can be used to determine the quark coupling parameters $\mathcal{A}_b$ and $\mathcal{A}_c$ and combined with the corresponding direct measurements at SLD. This is illustrated in Fig. 3.2, which shows the different measurements in the $(\mathcal{A}_b, \mathcal{A}_\ell)$ and $(\mathcal{A}_c, \mathcal{A}_\ell)$ planes together with the contours of the average and the theoretical prediction. The quark coupling parameters are very insensitive to radiative

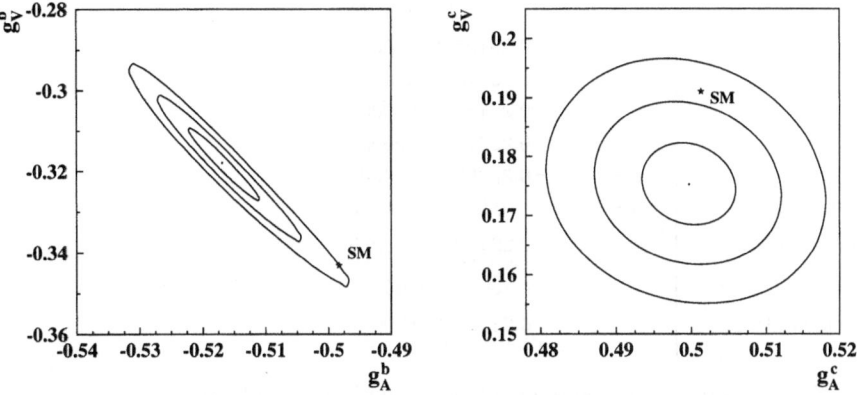

Fig. 3.3. Fit contours (one, two and three sigma) of g_V^b versus g_A^b (*left*) and g_V^c versus g_A^c (*right*). The *star* indicates the Standard Model prediction

corrections, in contrast to the strong dependence of $\mathcal{A}_\ell$ on the Standard Model input parameters.

The results for both c and b quark couplings are low compared to the Standard Model prediction.

$$\mathcal{A}_c = 0.625 \pm 0.021 \;,$$

is about 1.9 sigma below the prediction of 0.6656 ± 0.0022 and

$$\mathcal{A}_b = 0.892 \pm 0.016$$

is 2.6 sigma away from the Standard Model value of 0.934 ± 0.0004. Part of these descrepancies can be attributed to the large measured value for $\mathcal{A}_\ell$, since the $\mathcal{A}_q$ determined from the forward–backward asymmetries at LEP are proportional to $1/\mathcal{A}_\ell$. The low value for $A_{\mathrm{FB}}^{0,b}$ at LEP and $\mathcal{A}_c$ at SLD enhances the effect.

3.1.3 Heavy-Quark Couplings

The couplings g_V and g_A for b and c quarks can be determined from the partial widths and asymmetry measurements at LEP and SLD in a similar manner to the leptonic couplings. However, for quarks additional complications arise owing to QCD corrections and mass effects. Beyond one-loop order mixed terms occur which involve electroweak parameters such as m_t and m_H and the QCD coupling α_{s} (see Sect. 1.4.3). Therefore, in the case of quarks one cannot unambiguously define effective couplings which contain only electroweak corrections. The results presented here follow the conventions used in ZFITTER [101], where these terms are factorized.

Figure 3.3 shows the contours of g_V and g_A for heavy quarks. A striking feature is the high correlation (98 %) of the b quark couplings. This is due

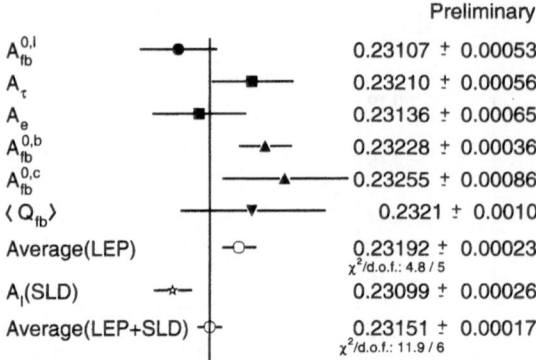

Fig. 3.4. Asymmetry measurements expressed in terms of $\sin^2\theta_{\text{eff}}^{\text{lept}}$

to the combination of two effects. Unlike the case of leptons, there is no decoupling of the quark width and asymmetry measurements into g_A and g_V, since g_V and g_A are numerically similar, in particular for down-type quarks. In addition, for b quarks the partial width (R_b) is known very accurately, which constrains g_V^b and g_A^b onto a small ring around the origin. The b asymmetries give a perpendicular but much weaker constraint. Together this leads to the strong correlation. For c quarks the accuracies of the partial width and asymmetry are relatively similar when projected into the (g_V, g_A) plane and the correlation is largely compensated.

The results for the c and b effective couplings deviate from the Standard Model prediction by 2.3 and 3 sigma, respectively. Most of this difference can be attributed to the descrepancies in $\mathcal{A}_c$ and $\mathcal{A}_b$ discussed above. The small deviations in R_c and R_b further enhance the difference.

3.1.4 Electroweak Mixing Angle

With the assumption of lepton universality for the leptonic asymmetries (forward–backward, τ polarization and left–right asymmetries) and using the Standard Model predictions for the quark couplings, which are very insensitive to the Standard Model parameters, one can express all[3] asymmetry and polarization measurements in terms of the effective leptonic electroweak mixing angle (1.124)

$$\sin^2\theta_{\text{eff}}^{\text{lept}} \equiv \frac{1}{4}\left(1 - \frac{g_V^\ell}{g_A^\ell}\right).$$

Figure 3.4 summarizes these measurements in terms of $\sin^2\theta_{\text{eff}}^{\text{lept}}$. The average is

$$\sin^2\theta_{\text{eff}}^{\text{lept}} = 0.23151 \pm 0.00017 \tag{3.1}$$

---

[3] Except $\mathcal{A}_c$ and $\mathcal{A}_b$ obtained from SLD, which determine the final-state couplings.

with a χ^2/d.o.f. of 11.9/6, corresponding to a probability of 6%. The largest discrepancy is found between the two most precise determinations, A_{LR} and $A_{\mathrm{FB}}^{0,b}$, which differ by 2.9 standard deviations. This discrepancy has the same origin as the $\mathcal{A}_b$ discrepancy (Sect. 3.1.2), but now it is projected on to $\sin^2\theta_{\mathrm{eff}}^{\mathrm{lept}}$, since Standard Model quark couplings are assumed for $A_{\mathrm{FB}}^{0,b}$.

3.2 Effect of Radiative Corrections

As discussed in Sects. 1.3 and 1.4, electroweak radiative corrections modify the tree-level predictions for the couplings and the vector boson mass relations. One can divide these into four categories and the corresponding experimental observables:

- the Z^0 decay widths, in particular the leptonic widths, are affected predominantly by the normalization in terms of ρ_f (1.119);
- the asymmetry measurements expressed as $\sin^2\theta_{\mathrm{eff}}^{\mathrm{lept}}$ are sensitive to κ (1.119);
- R_b, the ratio of the b width to the total hadronic width, measures the b-specific corrections ρ_b (1.125);
- The W–Z mass relation depends on Δr (1.113).

The experimental sensitivity to radiative corrections is demonstrated in Fig. 3.5, which shows the experimental results for Γ_ℓ, $\sin^2\theta_{\mathrm{eff}}^{\mathrm{lept}}$, R_b and m_W/m_Z together with the Standard Model predictions as a function of m_H and the tree-level expectations, which account only for the energy dependence of α (Sect. 1.2.3).

In particular for Γ_ℓ and m_W/m_Z the effects of radiative corrections are large and unambiguously established by the measurements. The value of $\sin^2\theta_{\mathrm{eff}}^{\mathrm{lept}}$ varies most strongly with m_H. However, the theoretical prediction suffers from an additional dependence on the electromagnetic coupling constant $\alpha(m_Z^2)$ (Sect. 1.2.3), which induces a larger uncertainty (0.00023) in $\sin^2\theta_{\mathrm{eff}}^{\mathrm{lept}}$ than the experimental error (0.00017). R_b is essentially only sensitive to m_t via vertex corrections, since m_t- and m_H-dependent propagator corrections cancel in the ratio of partial widths.

3.2.1 Epsilon Parameters

An alternative approach to separating the radiative corrections is provided by the ϵ parameters, described in Sect. 1.5.1 and [110]. The ϵ parameters combine the aforementioned parameters ρ, κ, Δr and ρ_b in such a way that large corrections quadratic in m_t are absorbed into ϵ_1 and ϵ_b; ϵ_2 and ϵ_3 have only a small residual logarithmic dependence on m_t, and the leading logarithmic m_H-dependent terms affect only ϵ_1 and ϵ_3.

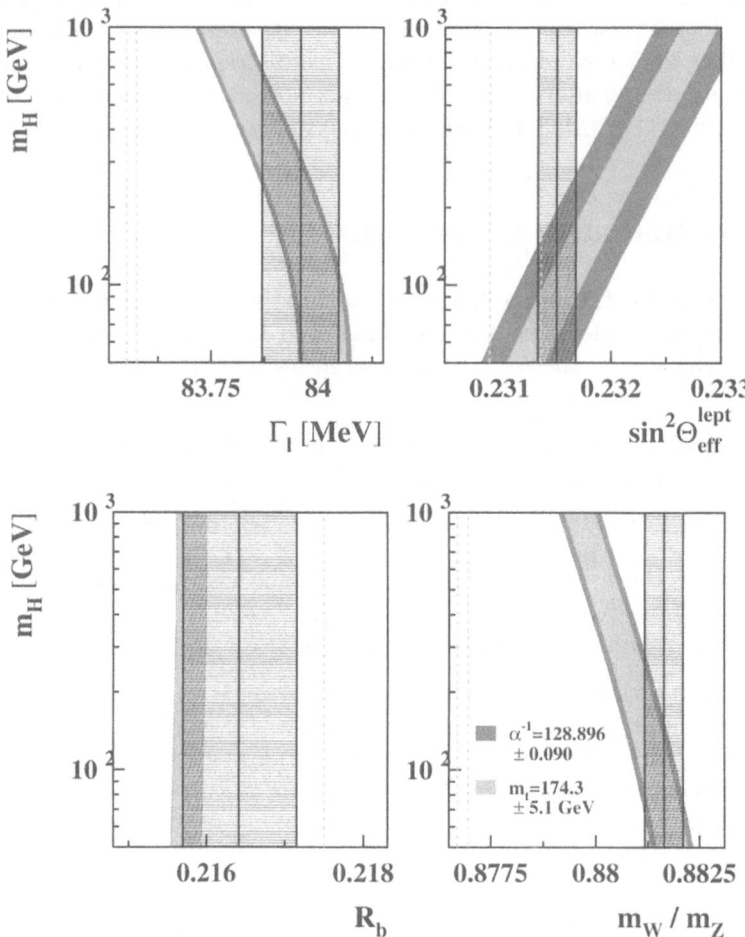

Fig. 3.5. Effect of radiative corrections for Γ_ℓ (*top left*), $\sin^2\theta_{\text{eff}}^{\text{lept}}$ (*top right*), R_b (*bottom left*) and m_W/m_Z (*bottom right*). The sloping band in each panel shows the Standard Model prediction as a function of m_H and the dependence on m_t and α. Also shown are the one-sigma bands of the experimental results (*horizontally hatched*) and the tree-level predictions (*vertical dashed lines*)

The ϵ parameters directly quantify the electroweak corrections since they vanish at tree-level. Furthermore they separate the corrections according to their dependence on unknown or weakly constrained Standard Model parameters. "New physics" affecting the gauge boson self-energies or the b couplings would show up more directly in the ϵ parameters.

Using the electroweak measurements, the four ϵ parameters can be determined in a fit together with m_Z, α_s and α. Since m_t cannot be expressed in terms of these parameters the measurements of m_t (TEVATRON) and

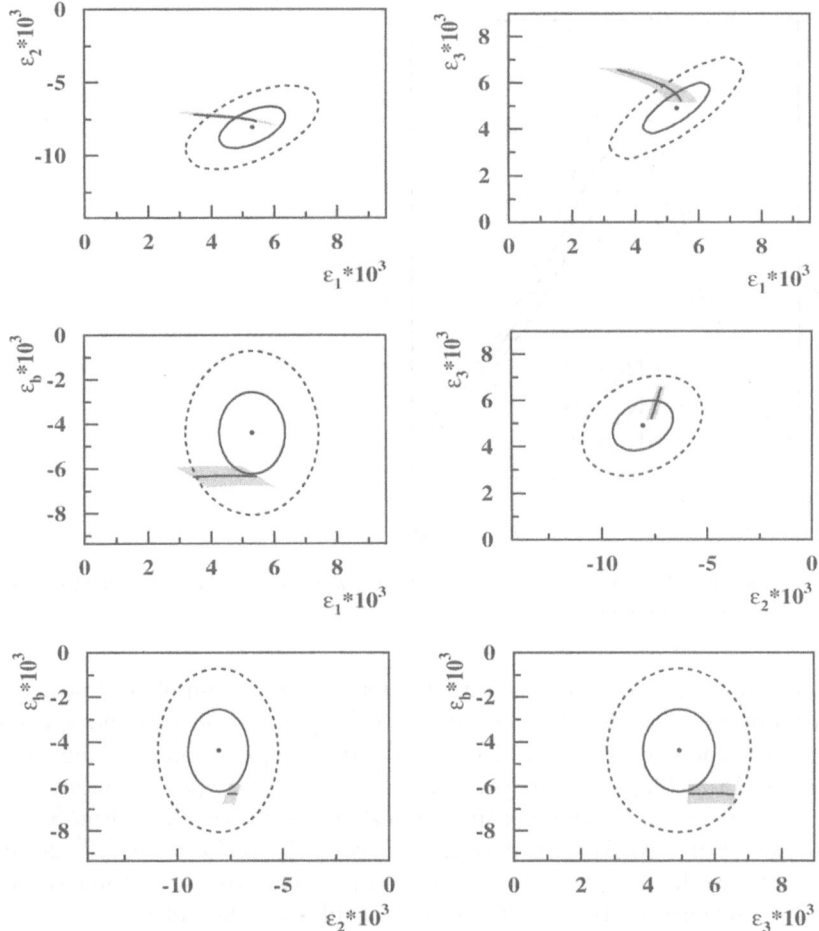

Fig. 3.6. One and two sigma contour plots for the ϵ parameters. The shaded areas show the Standard Model predictions (conventions as in Fig. 3.1)

$\sin^2\theta_W$ (NUTEV) are not used in the fit. Figure 3.6 shows the contour plots for the ϵ parameters. The values of ϵ_1, ϵ_2, ϵ_3 and ϵ_b agree within one sigma with the Standard Model predictions. The results once more demonstrate the presence of quantum loop corrections; at tree-level all ϵ parameters would be zero.

3.3 Number of Light Neutrino Species

One of the most fundamental questions in particle physics is the number of fermion family generations, which is not constrained in the Standard Model. With the precise measurement of the Z^0 resonance at LEP this parameter

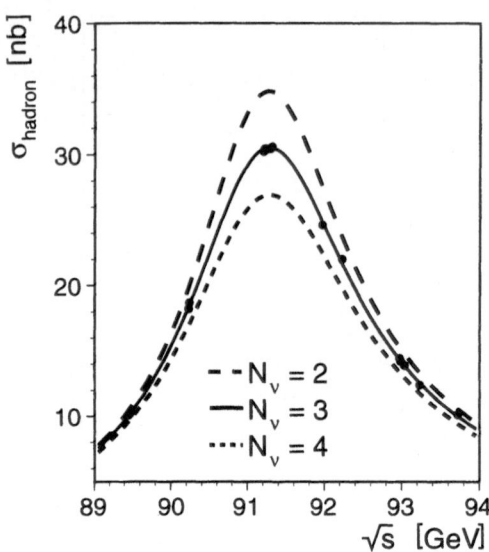

Fig. 3.7. Hadronic lineshape measured at LEP *(points)* and the expected curves for 2, 3 and 4 neutrino species

can be determined. The total Z^0 width Γ_Z depends on the number of possible decay channels and is therefore sensitive to the number of neutrino species N_ν. Similarly, the hadronic pole cross-section σ_{had}^0 depends on N_ν, since it is a product of branching ratios ($\sigma_{\text{had}}^0 \propto Br(Z^0 \to e^+e^-)\,Br(Z^0 \to \text{hadrons})$). The large effect of N_ν on the measured lineshape is illustrated in Fig. 3.7.

Experimentally, N_ν is best determined from the ratio of the Z^0 decay width into invisible particles ($\Gamma_{\text{inv}} \equiv \Gamma_Z - \Gamma_{\text{had}} - 3\,\Gamma_\ell$) to the leptonic decay width. By transforming the results given in Table 2.6, one obtains

$$\Gamma_{\text{inv}}/\Gamma_\ell = 5.9407 \pm 0.016 \ . \tag{3.2}$$

This quantity has the advantage, that the theoretical uncertainties induced by radiative corrections and by the limited knowledge of Standard Model parameters largely cancel.

With the Standard Model value for the ratio of the neutrino to the lepton partial width, $(\Gamma_\nu/\Gamma_\ell)_{\text{SM}} = 1.9912 \pm 0.0012$, N_ν can be calculated from the ratio

$$N_\nu = \frac{\Gamma_{\text{inv}}/\Gamma_\ell}{(\Gamma_\nu/\Gamma_\ell)_{\text{SM}}} = 2.9835 \pm 0.0085 \ , \tag{3.3}$$

which is consistent with three neutrino species. Within the Standard Model neutrinos are massless and the number of fermion generations is unambiguously determined.

Of course, one can turn the argument around and assume three neutrino generations with identical couplings and the Standard Model prediction $g_V^\nu \equiv g_A^\nu$ to obtain a precision measurement of the neutrino coupling:

$g^{\nu} = 0.50058 \pm 0.00075$.

If in addition the Standard Model couplings for the neutrinos are used, one can constrain the Z width from additional invisible decays of the Z^0:

$\Delta\Gamma_{\mathrm{inv}} = -2.9 \pm 1.7\,\mathrm{MeV}$.

In extensions of the Standard Model additional particles could be produced in Z^0 decays and contribute to the observed invisible width. This determination of $\Delta\Gamma_{\mathrm{inv}}$ sets significant constraints on such models (see [217] for an example).

3.4 Electroweak Fits

Finally, the electroweak precision measurements are used here to fit directly the parameters of the Standard Model. The fits are based on the packages ZFITTER [101] and TOPAZ [102], which incorporate recent theoretical calculations as discussed in Sect. 1.4.5.

The following set of Standard Model parameters was determined in the fits:

- mass of the Z^0 Boson, m_Z;
- mass of the top quark, m_t;
- mass of the Higgs boson, m_H;
- strong coupling constant, $\alpha_s(m_Z^2)$;
- electromagnetic coupling constant, $\alpha(m_Z^2)$.

From these parameters other variables of interest, such as m_W or $\sin^2\theta_{\mathrm{eff}}^{\mathrm{lept}}$, can be derived.

As an additional input the Fermi constant G_μ (1.83) was used, which relates the vector boson masses m_W and m_Z and the electromagnetic coupling constant α (1.34). Its uncertainty is below 10 ppm and negligible for the analysis.

The hadronic vacuum polarization (Sect. 1.2.3) of the photon causes large uncertainties for the evolution of the electromagnetic coupling constant from 0 to m_Z. The central value used for the analysis was

$$1/\alpha(m_Z^2) = 128.878 \pm 0.090 , \tag{3.4}$$

based on the evaluation in [35].

Four different fits were performed, including and excluding the direct measurements of m_t and m_W. The comparison of the indirect results for m_W and m_t derived from the fit with the direct measurements scrutinizes the consistency of the Standard Model. The results of these fits are presented in Table 3.2.

First, the direct measurements of m_t and m_W were both excluded. The fitted values for m_W and m_t (Table 3.2, column 2) are a little lower than the

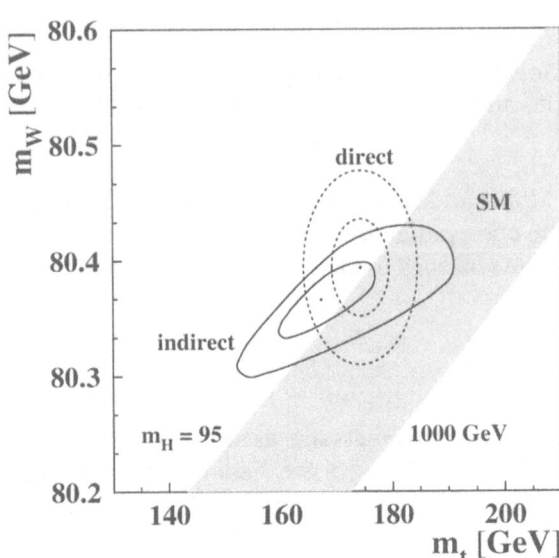

Fig. 3.8. One- and two-sigma contours in the (m_W, m_t) plane for the direct measurements (*dashed*) and for values obtained indirectly from a fit excluding the direct measurements (*solid*). The *band* shows the Standard Model prediction

Table 3.2. Standard Model fits to electroweak precision measurements, excluding and including the direct m_W and m_t results; m_W is derived from the fitted values

	No direct m_W and m_t	No direct m_t	No direct m_W	All data
m_t (GeV)	$167.3^{+10.5}_{-8.3}$	$169.7^{+9.8}_{-7.0}$	172.9 ± 4.7	173.2 ± 4.5
m_H (GeV)	55^{+84}_{-27}	57^{+93}_{-30}	81^{+77}_{-42}	77^{+69}_{-39}
$\alpha_s(m_Z^2)$	0.1183 ± 0.0026	0.1183 ± 0.0026	0.1185 ± 0.0026	0.1184 ± 0.0026
m_W (GeV)	80.366 ± 0.035	80.378 ± 0.026	80.381 ± 0.026	80.385 ± 0.022

direct measurements, but consistent within the uncertainties, as illustrated in Fig. 3.8.

In the second fit the direct m_W measurements were included to obtain the best indirect determination of m_t. This yields $m_t = 169.7^{+9.8}_{-7.0}$ GeV, in excellent agreement with the direct measurement of 174.3 ± 5.1 GeV.

For the third fit the approach was reversed; m_t was included and the m_W measurements were excluded to obtain the most precise indirect determination of $m_W = 80.381 \pm 0.026$ GeV, which is compatible with the average of the direct measurements of 80.394 ± 0.042 GeV.

Finally, the full set of measurements was used in the fit. The predictions of the Standard Model according to this fit for each observable, and the pulls,

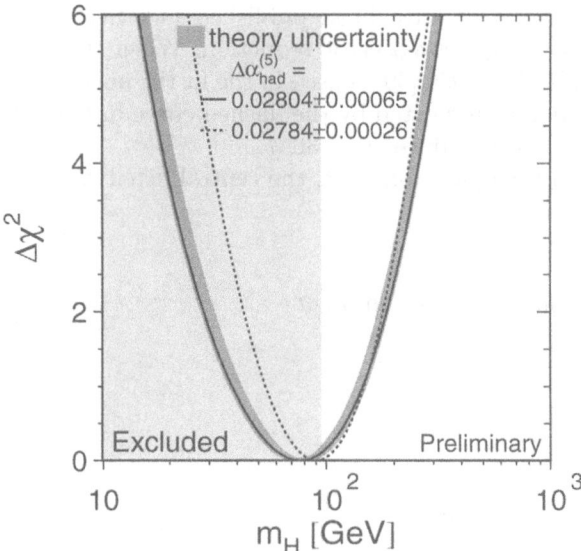

Fig. 3.9. The change of χ^2 as a function of m_H. The *line* is the result of the fit using all measurements and the *band* represents an estimate of theoretical uncertainties due to missing higher-order corrections. From this curve a 95 % upper limit for m_H of 215 GeV can be derived, which corresponds to a change in χ^2 of 2.7. Also shown is the result of a fit using a more precise value of $\alpha(m_Z^2)$ and the 95 % exclusion limit obtained from direct searches

are listed in Table 3.1. The χ^2/d.o.f. of the fit is 23/15, which corresponds to a probability of 8.6 %. Three measurements deviate by about two standard deviations: the b quark forward–backward asymmetry at LEP, the effective electroweak mixing angle determined from A_{LR} at SLD and the hadronic pole cross-section at LEP.

The value determined for the strong coupling, $\alpha_{\mathrm{s}}(m_Z^2) = 0.1184 \pm 0.0026$, is in perfect agreement with the world average of $\alpha_{\mathrm{s}}(m_Z^2) = 0.119 \pm 0.002$ [104].

Figure 3.9 shows the $\Delta\chi^2$ curve as a function of m_H. The width of the band indicates theoretical uncertainties due to missing higher-order radiative corrections estimated with ZFITTER and TOPAZ [103]. They contribute only marginally to the total uncertainty of m_H, compared with the experimental errors and the uncertainty of $\alpha(m_Z^2)$. From this curve an upper limit at the 95 % confidence level can be derived,

$$m_H < 215 \text{ GeV}, \tag{3.5}$$

which includes theoretical uncertainties. The lower limit on m_H of 95 GeV from direct searches did not enter the determination of this limit.

The precision obtainable for m_H is severely compromised by the uncertainty in $\alpha(m_Z^2)$; the fitted values for m_H and α have a correlation of -69%.

Recently, new evaluations of $\alpha(m_Z^2)$ have been published (see Sect. 1.2.3 and [39,41]) which lead to a theory-driven error reduction. When the more precise value $1/\alpha(m_Z^2) = 128.905 \pm 0.036$ [39] is used in the fit the uncertainty in m_H is significantly improved, as indicated by the dashed curve in Fig. 3.9. The correlation between m_H and α reduces to -36%.

Since the value for $\alpha(m_Z^2)$ is slightly different, the central fitted value for m_H also moves:

$$m_H = 90^{+57}_{-37} \text{ GeV} \;. \tag{3.6}$$

The 95% confidence level limit reduces to 200 GeV.

4. Conclusions

Experiments have measured the electroweak observables of the W and Z^0 bosons and the top quark with an impressive accuracy, thus providing precision tests of the electroweak theory at a level which had never been reached before.

For the interpretation of the precision experiments radiative corrections, better referred to as quantum effects, play a crucial role. The calculation of radiative corrections has been well established over the last few years, improving and stabilizing the Standard Model predictions. LEP has provided the mass of the Z boson as the third highly accurate input parameter, besides α and G_μ, for completion of the input; each other precision observable provides a test of the electroweak theory. The theoretical predictions of the Standard Model are sensitive to the mass of the top quark, now discovered, and to the as yet experimentally unknown Higgs boson through the virtual presence of these particles in the loops. As a consequence, precision data can be used to pin down the allowed ranges of the mass parameters, yielding m_t in beautiful agreement with the directly measured value. Although the dependence on the Higgs mass is only logarithmic, the experimental precision has reached a level where the sensitivity to m_H becomes visible, with a preference for a light Higgs well below the non-perturbative regime.

The overall agreement between theory and experiment for the entire set of precision observables is impressive. However, the two most precise measurements of the effective leptonic weak mixing angle $\sin^2\theta_{\text{eff}}^{\text{lept}}$ – the left-right cross-section asymmetry A_{LR} at SLD and the b quark forward–backward asymmetry $A_{\text{FB}}^{0,b}$ – differ by 2.9 standard deviations. Whether this is an experimental problem, an extreme statistical fluctuation or a real effect in the b couplings will presumably remain unsolved in the foreseeable future, since both LEP1 and SLC finished their operation and the full data sets have been analyzed, although some results are still preliminary.

Progress in electroweak precision measurements in the near future can be expected from several places. LEP2 will continue data-taking until the end of 2000 at a centre-of-mass energy slightly above to 200 GeV. The projected goal for m_W is to reduce the uncertainty to 30 MeV. More precise results for m_W and m_t are also expected from the TEVATRON run II, which starts in the year 2000. The experiments CDF and D0 aim to reach a 2 GeV precision for m_t and 30 MeV for m_W. After that the experiments at the new proton–proton collider LHC at CERN are scheduled to begin data-taking in 2005 and should further improve these measurements. Important experimental input for the hadronic vacuum polarization, which limits the accuracy of $\alpha(m_Z^2)$, is expected from the BES experiment [218] at the BEPC collider. More precise measurements of the hadronic cross-section at energies around 3 GeV will improve the uncertainty of the $\alpha(m_Z^2)$ determination currently used for the electroweak fit results and test some of the theoretical assumptions made for the recent more precise $\alpha(m_Z^2)$ evaluations.

In order to improve the indirect Higgs search also a further reduction of the theoretical uncertainties is important. The uncertainties in the Standard Model predictions have their origin essentially in the uncertainties of the hadronic vacuum polarization of the photon and in the unknown higher-order contributions. In order to reach a theoretical accuracy safely below the 0.1% level, new experimental data on $\Delta\alpha$ and more complete electroweak two-loop calculations are required. The minimal supersymmetric Standard Model (MSSM) describes the experimental data as good as the Standard Model. The W mass predicted by the minimal supersymmetric Standard Model (MSSM) is always higher than in the Standard Model; from the increasing experimental accuracy of m_W in the future one hence expects an important probe of both the Standard Model and the MSSM.

The crucial task in particle physics remains the search for the Higgs boson. At LEP2 the Standard Model Higgs boson can either be found or excluded up to $m_H \leq 110\,\mathrm{GeV}$. The TEVATRON run II will presumably push the limit further up to about 130 GeV. In case it still eludes detection, the LHC experiments will eventually cover the whole mass range up to 1 TeV. The energy range, which will be explored by LHC, should be sufficient to find either ultimately the Higgs boson or other new particles predicted in extensions of the Standard Model, such as supersymmetric particles or additional vector bosons.

References

1. S.L. Glashow, *Nucl. Phys.* B **22** (1961) 579; S. Weinberg, *Phys. Rev. Lett.* **19** (1967) 1264; A. Salam, in: *Proceedings of the 8th Nobel Symposium*, p. 367, ed. N. Svartholm, Almqvist and Wiksell, Stockholm 1968
2. S.L. Glashow, I. Iliopoulos, L. Maiani, *Phys. Rev.* D **2** (1970) 1285
3. N. Cabibbo, *Phys. Rev. Lett.* **10** (1963) 531; M. Kobayashi, K. Maskawa, *Prog. Theor. Phys.* **49** (1973) 652
4. H.Y. Han, Y. Nambu, *Phys. Rev.* **139** (1965) 1006; C. Bouchiat, I. Iliopoulos, Ph. Meyer, *Phys. Lett.* B **138** (1972) 652
5. G. 't Hooft, *Nucl. Phys.* B **33** (1971) 173; *Nucl. Phys.* B **35** (1971) 167
6. CDF Collaboration, F. Abe et al., *Phys. Rev. Lett.* **74** (1995) 2626; D0 Collaboration, S. Abachi et al., *Phys. Rev. Lett.* **74** (1995) 2632
7. L.D. Faddeev, V.N. Popov, *Phys. Lett.* B **25** (1967) 29
8. M. Böhm, W. Hollik, H. Spiesberger, *Fortschr. Phys.* **34** (1986) 687
9. G. Passarino, in: *Proceedings of the LP-HEP 91 Conference*, Geneva 1991, eds. S. Hegarty, K. Potter, E. Quercigh, World Scientific, Singapore 1992
10. G. Passarino, *Phys. Lett.* B **231** (1989) 458; *Phys. Lett.* B **247** (1990) 587; *Nucl. Phys.* B **361** (1991) 351
11. G. 't Hooft, *Nucl. Phys.* B **61** (1973) 455; *Nucl. Phys.* B **62** (1973) 444
12. D.A. Ross, J.C. Taylor, *Nucl. Phys.* B **51** (1973) 25
13. G. Passarino, M. Veltman, *Nucl. Phys.* B **160** (1979) 151
14. M. Consoli, *Nucl. Phys.* B **160** (1979) 208
15. A. Sirlin, *Phys. Rev.* D **22** (1980) 971; W.J. Marciano, A. Sirlin, *Phys. Rev.* D **22** (1980) 2695; A. Sirlin, W.J. Marciano, *Nucl. Phys.* B **189** (1981) 442
16. D.Yu. Bardin, P.Ch. Christova, O.M. Fedorenko, *Nucl. Phys.* B **175** (1980) 435; *Nucl. Phys.* B **197** (1982) 1; D.Yu. Bardin, M.S. Bilenky, G.V. Mithselmakher, T. Riemann, M. Sachwitz, *Z. Phys.* C **44** (1989) 493
17. J. Fleischer, F. Jegerlehner, *Phys. Rev.* D **23** (1981) 2001
18. K.I. Aoki, Z. Hioki, R. Kawabe, M. Konuma, T. Muta, *Suppl. Prog. Theor. Phys.* **73** (1982) 1; Z. Hioki, *Phys. Rev. Lett.* **65** (1990) 683, Erratum: *Phys. Rev. Lett.* **65** (1990) 1692; *Z. Phys.* C **49** (1991) 287
19. M. Consoli, S. LoPresti, L. Maiani, *Nucl. Phys.* B **223** (1983) 474
20. D.Yu. Bardin, M.S. Bilenky, G.V. Mithselmakher, T. Riemann, M. Sachwitz, *Z. Phys.* C **44** (1989) 493
21. W. Hollik, *Fortschr. Phys.* **38** (1990) 165; *Precision Tests of the Standard Model*, pp. 37 and 117, ed. P. Langacker, World Scientific, Singapore 1995
22. M. Consoli, W. Hollik, F. Jegerlehner, in: *Z Physics at LEP 1*, eds. G. Altarelli, R. Kleiss, C. Verzegnassi, CERN 89-08 (1989)
23. G. Passarino, R. Pittau, *Phys. Lett.* B **228** (1989) 89; V.A. Novikov, L.B. Okun, M.I. Vysotsky, *Nucl. Phys.* B **397** (1993) 35

24. M. Veltman, *Phys. Lett.* B **91** (1980) 95; M. Green, M. Veltman, *Nucl. Phys.* B **169** (1980) 137, Erratum: *Nucl. Phys.* B **175** (1980) 547; F. Antonelli, M. Consoli, G. Corbo, *Phys. Lett.* B **91** (1980) 90; F. Antonelli, M. Consoli, G. Corbo, O. Pellegrino, *Nucl. Phys.* B **183** (1981) 195
25. G. Passarino, M. Veltman, *Phys. Lett.* B **237** (1990) 537
26. W.J. Marciano, A. Sirlin, *Phys. Rev. Lett.* **46** (1981) 163; A. Sirlin, *Phys. Lett.* B **232** (1989) 123
27. G. Degrassi, S. Fanchiotti, A. Sirlin, *Nucl. Phys.* B **351** (1991) 49
28. G. Degrassi, A. Sirlin, *Nucl. Phys.* B **352** (1991) 342
29. D.C. Kennedy, B.W. Lynn, *Nucl. Phys.* B **322** (1989) 1
30. M. Kuroda, G. Moultaka, D. Schildknecht, *Nucl. Phys.* B **350** (1991) 25
31. A. Denner, T. Sack, *Nucl. Phys.* B **347** (1990) 203
32. J.C. Ward, *Phys. Rev.* **78** (1950) 1824
33. G. Källén, A. Sabry, *K. Dan. Vidensk. Selsk. Mat.-Fys. Medd.* **29** (1955) No. 17
34. M. Steinhauser, *Phys. Lett.* B **429** (1998) 158
35. S. Eidelman, F. Jegerlehner, *Z. Phys.* C **67** (1995) 585
36. H. Burkhardt, B. Pietrzyk, *Phys. Lett.* B **356** (1995) 398
37. M.L. Swartz, *Phys. Rev.* D **53** (1996) 5268
38. R. Alemany, M. Davier, A. Höcker, *Eur. Phys. J.* C **2** (1998) 123
39. M. Davier, A. Höcker, *Phys. Lett.* B **419** (1998) 419; hep-ph/9801361; *Phys. Lett.* B **435** (1998) 427
40. A. Höcker, in: *Proceedings of the XXIX International Conference on High Energy Physics*, Vancouver 1998, eds. A. Astbury, D. Axen, J. Robinson, World Scientific, Singapore 1999
41. J.H. Kühn, M. Steinhauser, *Phys. Lett.* B **437** (1998) 425
42. S. Groote, J. Körner, K. Schilcher, N.F. Nasrallah, *Phys. Lett.* B **440** (1998) 375
43. J. Erler, *Phys. Rev.* D **59** (1999) 054008
44. A.H. Hoang, J.H. Kühn, T. Teubner, *Nucl. Phys.* B **452** (1995) 173; K.G. Chetyrkin, J.H. Kühn, M. Steinhauser, *Phys. Lett.* B **371** (1996) 93; *Nucl. Phys.* B **482** (1996) 213; *Nucl. Phys.* B **505** (1997) 40; K.G. Chetyrkin, R. Harlander, J.H. Kühn, M. Steinhauser, *Nucl. Phys.* B **503** (1997) 339
45. A.D. Martin, D. Zeppenfeld, *Phys. Lett.* B **345** (1995) 558
46. E. Braaten, S. Narison, A. Pich, *Nucl. Phys.* B **373** (1992) 581
47. Z. Zhan, BES Collaboration, in: *Proceedings of the XXIX International Conference on High Energy Physics*, Vancouver 1998, eds. A. Astbury, D. Axen, J. Robinson, World Scientific, Singapore 1999
48. F. Jegerlehner, DESY 99-007, hep-ph/9901386
49. S. Eidelman, F. Jegerlehner, A. Kataev, O. Veretin, *Phys. Lett.* B **454** (1999) 369
50. C. Bollini, J. Giambiagi *Nuovo Cim.* B **12** (1972) 20; J. Ashmore, *Nuovo Cim. Lett.* **4** (1972) 289; G. 't Hooft, M. Veltman, *Nucl. Phys.* B **44** (1972) 189
51. D. Ross, M. Veltman, *Nucl. Phys.* B **95** (1975) 135
52. M. Veltman, *Nucl. Phys.* B **123** (1977) 89; M.S. Chanowitz, M.A. Furman, I. Hinchliffe, *Phys. Lett.* B **78** (1978) 285
53. M. Veltman, *Acta Phys. Polon.* B **8** (1977) 475.
54. R.E. Behrends, R.J. Finkelstein, A. Sirlin, *Phys. Rev.* **101** (1956) 866; T. Kinoshita, A. Sirlin, *Phys. Rev.* **113** (1959) 1652
55. T. van Ritbergen, R. Stuart, *Phys. Lett.* B **437** (1998) 201; *Phys. Rev. Lett.* **82** (1999) 488
56. W.J. Marciano, *Phys. Rev.* D **20** (1979) 274
57. M. Consoli, W. Hollik, F. Jegerlehner, *Phys. Lett.* B **227** (1989) 167

58. J.J. van der Bij, F. Hoogeveen, *Nucl. Phys.* B **283** (1987) 477
59. R. Barbieri, M. Beccaria, P. Ciafaloni, G. Curci, A. Vicere, *Phys. Lett.* B **288** (1992) 95; *Nucl. Phys.* B **409** (1993) 105; J. Fleischer, F. Jegerlehner, O.V. Tarasov, *Phys. Lett.* B **319** (1993) 249
60. A. Djouadi, C. Verzegnassi, *Phys. Lett.* B **195** (1987) 265
61. L. Avdeev, J. Fleischer, S.M. Mikhailov, O. Tarasov, *Phys. Lett.* B **336** (1994) 560, Erratum: *Phys. Lett.* B **349** (1995) 597; K.G. Chetyrkin, J.H. Kühn, M. Steinhauser, *Phys. Lett.* B **351** (1995) 331
62. A. Djouadi, *Nuovo Cim.* A **100** (1988) 357; D.Yu. Bardin, A.V. Chizhov, Dubna preprint E2-89-525 (1989); B.A. Kniehl, *Nucl. Phys.* B **347** (1990) 86; F. Halzen, B.A. Kniehl, *Nucl. Phys.* B **353** (1991) 567; A. Djouadi, P. Gambino, *Phys. Rev.* D **49** (1994) 3499
63. B.A. Kniehl, J.H. Kühn, R.G. Stuart, *Phys. Lett.* B **214** (1988) 621; B.A. Kniehl, A. Sirlin, *Nucl. Phys.* B **371** (1992) 141; *Phys. Rev.* D **47** (1993) 883; S. Fanchiotti, B.A. Kniehl, A. Sirlin, *Phys. Rev.* D **48** (1993) 307
64. K.G. Chetyrkin, J.H. Kühn, M. Steinhauser, *Phys. Rev. Lett.* **75** (1995) 3394
65. A. Sirlin, *Phys. Rev.* D **29** (1984) 89
66. G. Degrassi, P. Gambino, A. Vicini, *Phys. Lett.* B **383** (1996) 219; G. Degrassi, P. Gambino, A. Sirlin, *Phys. Lett.* B **394** (1997) 188; G. Degrassi, P. Gambino, M. Passera, A. Sirlin, B **418** (1998) 209
67. S. Bauberger, G. Weiglein, *Nucl. Instrum. Meth.* A **389** (1997) 318; *Phys. Lett.* B **419** (1997) 333
68. G. Weiglein, *Acta Phys. Polon.* B **29** (1998) 2735; hep-ph/9901317
69. A. Stremplat, Diploma Thesis (Karlsruhe 1998)
70. P. Gambino, A. Sirlin, *Phys. Rev. Lett.* **73** (1994) 621
71. G.L. Fogli and D. Haidt, *Z. Phys.* C **40** (1988) 379; H. Abramowicz et al., *Phys. Rev. Lett.* **57** (1986) 298; A. Blondel et al, *Z. Phys.* C **45** (1990) 361; CHARM Collaboration, J.V. Allaby et al., *Phys. Lett.* B **177** (1987) 446; *Z. Phys.* C **36** (1987) 611; CHARM-II Collaboration, D. Geiregat et al., *Phys. Lett.* B **247** (1990) 131; *Phys. Lett.* B **259** (1991) 499; C.G. Arroyo et al., *Phys. Rev. Lett.* **72**, 3452 (1994); CCFR/NuTev Collaboration, K.S. McFarland et al., *Eur. Phys. J.* C **1** (1998) 509
72. NuTeV Collaboration, T. Bolton, in: *Proceedings of the XXIX International Conference on High Energy Physics*, Vancouver 1998, eds. A. Astbury, D. Axen, J. Robinson, World Scientific, Singapore 1999
73. C.H. Llewellyn Smith, *Nucl. Phys.* B **228** (1983) 205
74. G. Degrassi, P. Gambino, hep-ph/9905472
75. A.A. Akhundov, D. Bardin, T. Riemann, *Nucl. Phys.* B **276** (1986) 1; W. Beenakker, W. Hollik, *Z. Phys.* C **40** (1988) 141; J. Bernabeu, A. Pich, A. Santamaria, *Phys. Lett.* B **200** (1988) 569; *Nucl. Phys.* B **363** (1991) 326
76. A. Denner, W. Hollik, B. Lampe, *Z. Phys.* C **60** (1993) 93
77. F.A. Berends et al., in: *Z Physics at LEP 1*, CERN 89-08 (1989), eds. G. Altarelli, R. Kleiss, C. Verzegnassi, Vol. I, p. 89; W. Beenakker, F.A. Berends, S.C. van der Marck, *Z. Phys.* C **46** (1990) 687
78. A. Borelli, M. Consoli, L. Maiani, R. Sisto, *Nucl. Phys.* B **333** (1990) 357
79. G. Burgers, F.A. Berends, W. Hollik, W.L. van Neerven, *Phys. Lett.* B **203** (1988) 177
80. D.Yu. Bardin, A. Leike, T. Riemann, M. Sachwitz, *Phys. Lett.* B **206** (1988) 539
81. G. Valencia, S. Willenbrock, *Phys. Lett.* B **259** (1991) 373; R.G. Stuart, *Phys. Lett.* B **272** (1991) 353; A. Sirlin, *Phys. Rev. Lett.* **67** (1991) 2127; *Phys. Lett.* B **267** (1991) 240

82. A. Leike, T. Riemann, J. Rose, *Phys. Lett.* B **273** (1991) 513; T. Riemann, *Phys. Lett.* B **293** (1992) 451

83. G. Burgers, F.A. Berends, W.L. van Neerven, *Nucl. Phys.* B **297** (1988) 429, Erratum: *Nucl. Phys.* B **304** (1988) 921

84. S. Jadach, M. Skrzypek, B.F.L. Ward, *Phys. Lett.* B **257** (1991) 173; S. Jadach, M. Skrzypek, *Z. Phys.* C **49** (1991) 548; G. Montagna, O. Nicrosini, F. Piccinini, *Phys. Lett.* B **406** (1997) 243

85. K.G. Chetyrkin, A.L. Kataev, F.V. Tkachov, *Phys. Lett.* B **85** (1979) 277; M. Dine, J. Sapirstein, *Phys. Rev. Lett.* **43** (1979) 668; W. Celmaster, R. Gonsalves, *Phys. Rev. Lett.* **44** (1980) 560; S.G. Gorishny, A.L. Kataev, S.A. Larin, *Phys. Lett.* B **259** (1991) 144; L.R. Surguladze, M.A. Samuel, *Phys. Rev. Lett.* **66** (1991) 560; A. Kataev, *Phys. Lett.* B **287** (1992) 209

86. K.G. Chetyrkin, J.H. Kühn, *Phys. Lett.* B **248** (1990) 359; *Phys. Lett.* B **406** (1997) 102; K.G. Chetyrkin, J.H. Kühn, A. Kwiatkowski, *Phys. Lett.* B **282** (1992) 221; K.G. Chetyrkin, A. Kwiatkowski, *Phys. Lett.* B **305** (1993) 285; *Phys. Lett.* B **319** (1993) 307

87. B.A. Kniehl, J.H. Kühn, *Phys. Lett.* B **224** (1990) 229; *Nucl. Phys.* B **329** (1990) 547; K.G. Chetyrkin, J.H. Kühn, *Phys. Lett.* B **307** (1993) 127; S. Larin, T. van Ritbergen, J.A.M. Vermaseren, *Phys. Lett.* B **320** (1994) 159; K.G. Chetyrkin, O.V. Tarasov, *Phys. Lett.* B **327** (1994) 114

88. K.G. Chetyrkin, J.H. Kühn, A. Kwiatkowski, in: *Reports of the Working Group on Precision Calculations for the Z Resonance*, p. 175, CERN 95-03 (1995), eds. D. Bardin, W. Hollik, G. Passarino; K.G. Chetyrkin, J.H. Kühn, A. Kwiatkowski, *Phys. Rep.* **277** (1996) 189

89. J. Fleischer, F. Jegerlehner, P. Rączka, O.V. Tarasov, *Phys. Lett.* B **293** (1992) 437; G. Buchalla, A.J. Buras, *Nucl. Phys.* B **398** (1993) 285; G. Degrassi, *Nucl. Phys.* B **407** (1993) 271; K.G. Chetyrkin, A. Kwiatkowski, M. Steinhauser, *Mod. Phys. Lett.* A **8** (1993) 2785

90. A. Kwiatkowski, M. Steinhauser, *Phys. Lett.* B **344** (1995) 359; S. Peris, A. Santamaria, *Nucl. Phys.* B **445** (1995) 252

91. R. Harlander, T. Seidensticker, M. Steinhauser, *Phys. Lett.* B **426** (1998) 125; J. Fleischer, F. Jegerlehner, M. Tentyukov, O. Veretin, *Phys. Lett.* B **459** (1999) 625

92. A. Czarnecki, J.H. Kühn, *Phys. Rev. Lett.* **77** (1996) 3955; Erratum: *Phys. Rev. Lett.* **80** (1998) 893

93. A. Hoang, J.H. Kühn, T. Teubner, *Nucl. Phys.* B **455** (1995) 3; *Nucl. Phys.* B **452** (1995) 173

94. M. Böhm, W. Hollik, *Nucl. Phys.* B **204** (1982) 45; *Z. Phys.* C **23** (1984) 31

95. S. Jadach, J.H. Kühn, R.G. Stuart, Z. Wąs, *Phys. Lett.* B **38** (1988) 609; J.H. Kühn, R.G. Stuart, *Phys. Lett.* B **200** (1988) 360

96. D. Bardin, M.S. Bilenky, A. Chizhov, A. Sazonov, Yu. Sedykh, T. Riemann, M. Sachwitz, *Phys. Lett.* B **229** (1989) 405; D. Bardin, M.S. Bilenky, A. Chizhov, A. Sazonov, O. Fedorenko, T. Riemann, M. Sachwitz, *Nucl. Phys.* B **351** (1991) 1

97. W. Beenakker, F.A. Berends, W.L. van Neerven, in: *Proceedings of the 1989 Ringberg Workshop on Radiative Corrections for e^+e^- Collisions*, p. 3, ed. J.H. Kühn, Springer, Berlin, Heidelberg 1989

98. M. Böhm, W. Hollik et al., in: *Z Physics at LEP 1*, CERN 89-08 (1989), eds. G. Altarelli, R. Kleiss, C. Verzegnassi, Vol. I, p. 203; W. Beenakker, F.A. Berends, S.C. van der Marck, *Phys. Lett.* B **252** (1990) 299

99. J. Jersak, E. Laerman, P.M. Zerwas, *Phys. Rev.* D **25** (1980) 1218; A. Djouadi, *Z. Phys.* C **39** (1988) 561; A. Djouadi, B. Lampe, P. Zerwas, *Z. Phys.* C **67**

(1995) 123; J.H. Kühn, P.Zerwas et al., in: *Z Physics at LEP 1*, CERN 89-08 (1989), eds. G. Altarelli, R. Kleiss, C. Verzegnassi, Vol. I, p. 267

100. G. Altarelli, B. Lampe, *Nucl. Phys. B* **391** (1993) 3; V. Ravindran, W. van Neerven, *Phys. Lett. B* **445** (1998) 214; S. Catani, M. Seymour, hep-ph/9905424

101. D. Bardin et al., hep-ph/9412201

102. G. Montagna, O. Nicrosini, F. Piccinini, G. Passarino, hep-ph/9804211

103. D. Bardin, G. Passarino, hep-ph/9803425; D. Bardin, M. Grünewald, G. Passarino, hep-ph/9902452

104. Particle Data Group, C. Caso et al., *Eur. Phys. J. C* **3** (1998) 1

105. D. Bardin et al., hep-ph/9709229, in: *Reports of the Working Group on Precision Calculations for the Z Resonance*, CERN 95-03 (1995), eds. D. Bardin, W. Hollik, G. Passarino, p. 7

106. P. Gambino, A. Sirlin, G. Weiglein, hep-ph/9903249, *JHEP* **04** (1999) 025

107. S. Jadach et al., in: *Reports of the Working Group on Precision Calculations for the Z Resonance*, CERN 95-03 (1995), eds. D. Bardin, W. Hollik, G. Passarino, p. 341; S. Jadach, O. Nicrosini, in: *Physics at LEP 2*, CERN 96-01 (1996), eds. G. Altarelli, T. Sjöstrand, F. Zwirner

108. G. Burgers, F. Jegerlehner, in: *Z Physics at LEP 1*, CERN 89-08 (1989), eds. G. Altarelli, R. Kleiss, C. Verzegnassi

109. M.E. Peskin, T. Takeuchi, *Phys. Rev. Lett.* **65** (1990) 964

110. G. Altarelli, R. Barbieri, *Phys. Lett. B* **253** (1991) 161; G. Altarelli, R. Barbieri, S. Jadach, *Nucl. Phys. B* **269** (1992) 3; Erratum: *Nucl. Phys. B* **276** (1992) 444

111. D.C. Kennedy, P. Langacker, *Phys. Rev. Lett.* **65** (1990) 2967

112. W.J. Marciano, J.L. Rosner, *Phys. Rev. Lett.* **65** (1990) 2963

113. B.W. Lynn, M.E. Peskin, R.G. Stuart, in: *Physics with LEP*, CERN 86-02 (1986), eds. J. Ellis and R. Peccei

114. R. Barbieri, M. Frigeni, F. Caravaglios, *Phys. Lett. B* **279** (1992) 169; V.A. Novikov, L.B. Okun, M.I. Vysotsky, *Mod. Phys. Lett. A* **8** (1993) 2529; M. Bilenky, K. Kolodziej, M. Kuroda, D. Schildknecht, *Phys. Lett. B* **319** (1993) 319; S. Dittmaier, D. Schildknecht, M. Kuroda, *Nucl. Phys. B* **448** (1995) 3

115. B. Holdom, J. Terning, *Phys. Lett. B* **247** (1990) 88; M. Golden, L. Randall, *Nucl. Phys. B* **361** (1991) 3; C. Roiesnel, T.N. Truong, *Phys. Lett. B* **256** (1991) 439

116. G. Altarelli, R. Barbieri, F. Caravaglios, *Nucl. Phys. B* **405** (1993) 3; *Phys. Lett. B* **349** (1995) 145

117. D. Toussaint, *Phys. Rev. D* **18** (1978) 1626; J.M Frere, J. Vermaseren, *Z. Phys. C* **19** (1983) 63

118. S. Bertolini, *Nucl. Phys. B* **272** (1986) 77; W. Hollik, *Z. Phys. C* **32** (1986) 291; *Z. Phys. C* **37** (1988) 569

119. A. Denner, R. Guth, J.H. Kühn, *Phys. Lett. B* **240** (1990) 438

120. A. Denner, R. Guth, W. Hollik, J.H. Kühn, *Z. Phys. C* **51** (1991) 695

121. B.W. Lynn, E. Nardi, *Nucl. Phys. B* **381** (1992) 467

122. T. Blank, W. Hollik, *Nucl. Phys. B* **514** (1998) 113

123. G. Altarelli et al., *Nucl. Phys. B* **342** (1990) 15; *Phys. Lett. B* **245** (1990) 669; M.C. Gonzalez-Gacia, J.W.F. Valle, *Phys. Lett. B* **259** (1991) 365; J. Layssac, F.M. Renard, C. Verzegnassi, *Z. Phys. C* **53** (1992) 97; F. del Aguila, W. Hollik, J.M. Moreno, M. Quirós, *Nucl. Phys. B* **372** (1992) 3

124. H.P. Nilles, *Phys. Rep.* **110** (1984) 1; H. Haber, G. Kane, *Phys. Rep.* **117** (1985) 75

125. M. Carena, J. Espinosa, M. Quiros, C. Wagner, *Phys. Lett.* B **355** (1995) 209; M. Carena, M. Quiros, C. Wagner, *Nucl. Phys.* B **461** (1996) 407; H. Haber, R. Hempfling, A. Hoang, *Z. Phys.* C **75** (1997) 539; S. Heinemeyer, W. Hollik, G. Weiglein, *Phys. Rev.* D **58** (1998) 091701; *Phys. Lett.* B **440** (1998) 296; *Eur. Phys. J.* C **9** (1999) 343

126. J. Rosiek, *Phys. Lett.* B **252** (1990) 135; M. Boulware, D. Finnell, *Phys. Rev.* D **44** (1991) 2054

127. G. Altarelli, R. Barbieri, F. Caravaglios, *Phys. Lett.* B **314** (1993) 357; C.S. Lee, B.Q. Hu, J.H. Yang, Z.Y. Fang, *J. Phys.* G **19** (1993) 13; Q. Hu, J.M. Yang, C.S. Li, *Commun. Theor. Phys.* **20** (1993) 213; J.D. Wells, C. Kolda, G.L. Kane, *Phys. Lett.* B **338** (1994) 219; G.L. Kane, R.G. Stuart, J.D. Wells, *Phys. Lett.* B **354** (1995) 350; M. Drees et al., *Phys. Rev.* D **54** (1996) 5598

128. P. Chankowski, A. Dabelstein, W. Hollik, W. Mösle, S. Pokorski, J. Rosiek, *Nucl. Phys.* B **417** (1994) 101; D. Garcia, J. Solà, *Mod. Phys. Lett.* A **9** (1994) 211

129. D. Garcia, R. Jiménez, J. Solà, *Phys. Lett.* B **347** (1995) 309, 321; D. Garcia, J. Solà, *Phys. Lett.* B **357** (1995) 349; A. Dabelstein, W. Hollik, W. Mösle, in: *Proceedings of the 1995 Ringberg Workshop on Perspectives for Electroweak Interactions in e^+e^- Collisions*, ed. B.A. Kniehl, World Scientific, Singapore 1995, p. 345; P. Chankowski, S. Pokorski, *Nucl. Phys.* B **475** (1996) 3; J. Bagger, K. Matchev, D. Pierce, R. Zhang, *Nucl. Phys.* B **491** (1997) 3

130. W. de Boer, A. Dabelstein, W. Hollik, W. Mösle, U. Schwickerath, *Z. Phys.* C **75** (1997) 627

131. A. Djouadi, P. Gambino, S. Heinemeyer, W. Hollik, C. Jünger, G. Weiglein, *Phys. Rev. Lett.* **78** (1997) 3626; *Phys. Rev.* D **57** (1998) 4179

132. J. Erler, D. Pierce, *Nucl. Phys.* B **526** (1998) 53

133. S. Myers, *The LEP Collider*, CERN 91-08 (1991)

134. ALEPH Collaboration, D. Decamp et al., *Nucl. Instrum. Meth.* A **294** (1990) 121; ALEPH Collaboration, D. Buskulic et al., *Nucl. Instrum. Meth.* A **360** (1995) 481

135. DELPHI Collaboration, P. Aarnio et al., *Nucl. Instrum. Meth.* A **303** (1991) 233; DELPHI Collaboration, P. Abreu et al., *Nucl. Instrum. Meth.* A **378** (1996) 57

136. L3 Collaboration, B. Adeva et al., *Nucl. Instrum. Meth.* A **289** (1990) 35

137. OPAL Collaboration, K. Ahmet et al., *Nucl. Instrum. Meth.* A **305** (1991) 275; P. Allport et al., *Nucl. Instrum. Meth.* A **324** (1993) 34

138. R. Brun et al., GEANT3 User's Guide, CERN DD/EE/84-1 (Revised) (1987)

139. LEP Energy Working Group, R. Assmann et al., *Eur. Phys. J.* C **6** (1999) 187; LEP Energy Working Group, R. Assmann et al., *Z. Phys.* C **66** (1995) 567; L. Arnaudon et al., *Z. Phys.* C **66** (1995) 45; L. Arnaudon et al., *Phys. Lett.* B **284** (1992) 431; L. Knudsen et al., *Phys. Lett.* B **270** (1991) 97

140. A.A. Sokolov, I.M. Ternov, *Sov. Phys. Dokl.* **8** (1964) 1203

141. L. Arnaudon et al., *Nucl. Instrum. Meth.* A **357** (1995) 249

142. D. Berede et al., *Nucl. Instrum. Meth.* A **365** (1995) 117

143. I.C. Brock et al., *Nucl. Instrum. Meth.* A **381** (1996) 236

144. OPAL Collaboration, S. Arcelli, in: *Proceedings of the XXIX International Conference on High Energy Physics*, Vancouver 1998, eds. A. Astbury, D. Axen, J. Robinson, World Scientific, Singapore 1999

145. DELPHI Collaboration, S.J. Alvsvaag et al., *Nucl. Phys.* B *Proc. Suppl.* **44** (1995) 116

146. *Z Physics at LEP 1*, CERN 89-08 (1989), eds. G. Altarelli, R. Kleiss, C. Verzegnassi, Vol. I

147. B.F.L. Ward, in: *Proceedings of the XXIX International Conference on High Energy Physics*, Vancouver 1998, eds. A. Astbury, D. Axen, J. Robinson, World Scientific, Singapore 1999; B.F.L. Ward, S. Jadach, M. Melles, S.A. Yost, *New Results on the Theoretical Precision of the LEP/SLC Luminosity*, hep-ph/9811245

148. S. Jadach et al., *Comput. Phys. Commun.* **102** (1997) 229; S. Jadach et al., *Comput. Phys. Commun.* **70** (1992) 305

149. M. Caffo, E. Remiddi, H. Czyż, *Int. J. Mod. Phys.* **C4** (1993) 591; M. Caffo, H. Czyż, E. Remiddi, *Nuovo Cim.* **105A** (1992) 277

150. M. Böhm, A. Denner, W. Hollik, *Nucl. Phys.* B **304** (1988) 687; F.A. Berends, R. Kleiss, W. Hollik, *Nucl. Phys.* B **304** (1988) 712

151. S. Jadach, W. Płaczek, B.F.L. Ward, *Phys. Lett.* B **390** (1997) 298

152. S. Jadach, B.F.L. Ward, Z. Wąs, *Comput. Phys. Commun.* **79** (1994) 503

153. T. Sjöstrand, *Comput. Phys. Commun.* **82** (1994) 74

154. G. Marchesini et al., hep-ph/9607393 (1996); G. Marchesini, B.R. Webber, *Nucl. Phys.* B **310** (1988) 461

155. LEP/SLD Electroweak and Heavy Flavour Working Groups, *A Combination of Preliminary Electroweak Measurement and Constraints on the Standard Model*, CERN-EP/99-15 (1999)

156. ALEPH Collaboration, D. Buskulic et al., *Z. Phys.* C **62** (1994) 539; ALEPH Collaboration, *Measurement of the Z Resonance Parameters at LEP*, CERN-EP/99-104 (1999)

157. DELPHI Collaboration, P. Abreu et al., *Nucl. Phys.* B **418** (1994) 403; DELPHI Collaboration, DELPHI Note 97-130 CONF 109, contributed paper to EPS-HEP-97, Jerusalem, EPS-463

158. L3 Collaboration, M. Acciarri et al., *Z. Phys.* C **62** (1994) 551; L3 Collaboration, *Preliminary L3 Results on Electroweak Parameters using 1990–96 Data*, L3 Note 2065, March 1997

159. OPAL Collaboration, R. Akers et al., *Z. Phys.* C **61** (1994) 19; OPAL Collaboration, *Precision Measurements of the Z^0 Lineshape and Lepton Asymmetry*, OPAL Physics Note PN358, contributed paper to ICHEP 98, Vancouver ICHEP'98 #229

160. W. Beenakker, F.A. Berends, S.C. van der Marck, *Nucl. Phys.* B **349** (1991) 323

161. G. Quast, Z Lineshape, Lepton Forward–Backward Asymmetries and Standard Model Fits, talk presented at the EPS-HEP-99, Tampere

162. OPAL Collaboration, G. Abbiendi et al., *Phys. Lett.* B **444** (1998) 539

163. ALEPH Collaboration, D. Buskulic, *Phys. Lett.* B **365** (1996) 437

164. ALEPH Collaboration, D. Buskulic et al., *Z. Phys.* C **69** (1996) 183; ALEPH Collaboration, *Measurement of the Tau Polarisation by ALEPH with the Full LEP I Data Sample*, ALEPH 98-067 CONF 98-037, contributed paper to ICHEP 98, Vancouver, ICHEP'98 #939

165. DELPHI Collaboration, P. Abreu et al., *Z. Phys.* C **67** (1995) 183; DELPHI Collaboration, *A Precise Measurement of the τ Polarisation at LEP1*, DELPHI 99-130 CONF 317, contributed paper to EPS-HEP-99, Tampere, 5-79

166. L3 Collaboration, M. Acciarri et al., *Phys. Lett.* B **429** (1998) 387

167. OPAL Collaboration, G. Alexander et al., *Z. Phys.* C **72** (1996) 365

168. SLD Collaboration, K. Abe et al., *Phys. Rev. Lett.* **73** (1994) 25

169. SLD Collaboration, J. Brau, Electroweak Precision Measurements with Leptons, invited talk at EPS-HEP-99, Tampere; SLD Collaboration, *An Improved Direct Measurement of Leptonic Coupling Asymmetries with Polarized Z's*, SLAC PUB 7878, contributed paper to ICHEP98, Vancouver, ICHEP'98 #362

170. OPAL Collaboration, K. Ackerstaff et al., *Z. Phys.* C **76** (1997) 387
171. B. Adeva et al., *Phys. Rev. Lett.* **51** (1983) 443
172. W. Bartel et al., *Phys. Lett.* B **146** (1984) 437
173. T. Behnke, D.G. Charlton, *Phys. Scripta* **52** (1995) 133
174. K. Mönig, *Rept. Prog. Phys.* **61** (1998) 999
175. ALEPH Collaboration, R. Barate et al., *Phys. Lett.* B **401** (1997) 150; ALEPH Collaboration, R. Barate et al., *Phys. Lett.* B **401** (1997) 163
176. DELPHI Collaboration, P. Abreu et al., *A Precise Measurement of the Partial Decay Width Ratio* $R_b = \Gamma_{b\bar{b}}/\Gamma_{had}$, CERN-EP/98-180, accepted by *Eur. Phys. J.* C
177. L3 Collab., M. Acciarri et al., *Measurement of R_b and $Br(b \rightarrow l\nu X)$ at LEP Using Double-Tag Methods*, CERN-EP/99-121, submitted to *Eur. Phys. J.* C
178. OPAL Collaboration, G. Abbiendi et al., *Eur. Phys. J.* C **8** (1999) 217
179. SLD Collaboration, *A Preliminary Measurement of R_b Using the Upgrade SLD Vertex Detector*, SLAC-PUB-7585, contributed paper to EPS-HEP-97, Jerusalem, EPS-118
180. ALEPH Collaboration, R. Barate et al., *Eur. Phys. J.* C **4** (1998) 557; ALEPH Collaboration, *Study of Charmed Hadron Production in Z Decays*, contributed paper to EPS-HEP-97, Jerusalem, EPS-623
181. DELPHI Collaboration, *Summary of R_c Measurements in DELPHI*, DELPHI 96-110 CONF 37, contributed paper to ICHEP96, Warsaw, PA01-060; DELPHI Collaboration, P. Abreu et al., *Measurement of the Z Partial Decay Width into $c\bar{c}$ and Multiplicity of Charm Quarks per b Decay*, CERN-EP/99-66, accepted by *Eur. Phys. J.* C
182. OPAL Collaboration, K. Ackerstaff et al., *Eur. Phys. J.* C **1** (1998) 439; OPAL Collaboration, G. Alexander et al., *Z. Phys.* C **72** (1996) 1
183. SLD Collaboration, *A Measurement of R_c with the SLD Detector*, SLAC PUB 7880, contributed paper to ICHEP98, Vancouver, ICHEP'98 #174
184. LEP Experiments: ALEPH, DELPHI, L3 and OPAL, *Nucl. Instrum. Meth.* A **378** (1996) 101; LEP Heavy Flavour Group, *Input Parameters for the LEP/SLD Electroweak Heavy Flavour Results for Summer 1998 Conferences*, LEPHF/98-01, http://www.cern.ch/LEPEWWG/heavy/lephf9801.ps.gz
185. OPAL Collaboration, R. Akers et al., *Phys. Lett.* B **353** (1995) 595; ALEPH Collaboration, R. Barate et al., *Phys. Lett.* B **434** (1998) 437; DELPHI Collaboration, P. Abreu et al., *Phys. Lett.* B **405** (1997) 202
186. D. Abbaneo et al., *Eur. Phys. J.* C **4** (1998) 185
187. ALEPH Collaboration, R. Barate et al., *Phys. Lett.* B **426** (1998) 217; ALEPH Collaboration, R. Barate et al., *Phys. Lett.* B **434** (1998) 415 ALEPH Collaboration, *Measurement of the b and c Forward–Backward Asymmetries using Leptons*, ALEPH 99-076 CONF 99-048, contributed paper to EPS-HEP-99, Tampere, 5-400
188. DELPHI Collaboration, P.Abreu et al., *Z. Phys.* C **65** (1995) 569; DELPHI Collaboration, *Measurement of the Forward–Backward Asymmetries of $e^+e^- \rightarrow Z \rightarrow b\bar{b}$ and $e^+e^- \rightarrow Z \rightarrow c\bar{c}$ Using Prompt Leptons*, DELPHI 98-143 CONF 204, contributed paper to ICHEP98, Vancouver, ICHEP'98 #124; DELPHI Collaboration, P.Abreu et al., *Eur. Phys. J.* C **9** (1999) 367; DELPHI Collaboration, P.Abreu et al., *Measurement of the Forward–Backward Asymmetry of c and b Quarks at the Z Pole Using Reconstructed D Mesons*, CERN-EP/99-07, accepted by *Eur. Phys. J.* C
189. L3 Collaboration, M. Acciarri et al., *Phys. Lett.* B **448** (1999) 152; L3 Collaboration, $A_{\text{FB}}^b \bar{b}$ *Using a Jet Charge Technique on 1994 Data*, L3 Note 2129

190. OPAL Collaboration, K.Ackerstaff et al., *Z. Phys.* C **75** (1997) 385; OPAL Collaboration, G. Alexander et al., *Z. Phys.* C **73** (1996) 379; OPAL Collaboration, G. Alexander et al., *Z. Phys.* C **70** (1996) 357; OPAL Collaboration, R. Akers et al., *Updated Measurement of the Heavy Quark Forward–Backward Asymmetries and Average b Mixing Using Leptons in Multihadronic Events*, OPAL Note PN226, contributed paper to ICHEP96, Warsaw, PA05-007

191. SLD Collaboration, *Combined SLD Measurement of A_b using Various Techniques*, SLAC PUB 8211, contributed paper to EPS-HEP-99, Tampere, 6-473; SLD Collaboration, *Direct Measurement of A_c using Inclusive Charm Tagging at the SLD Detector*, SLAC PUB 8195, contributed paper to EPS-HEP-99, Tampere, 6-474; SLD Collaboration, *Measurement of A_c with Charmed Mesons at SLD*, SLAC PUB 8199, contributed paper to EPS-HEP-99, Tampere, 6-474

192. ALEPH Collaboration, D. Buskulic et al., *Z. Phys.* C **71** (1996) 357

193. DELPHI Collaboration, P. Abreu et al., *Phys. Lett.* B **277** (1992) 371; DELPHI Collaboration, *Measurement of the Inclusive Charge Flow in Hadronic Z Decays*, DELPHI 96-19 PHYS 594

194. L3 Collaboration, M. Acciarri et al., *Phys. Lett.* B **439** (1998) 225

195. OPAL Collaboration, P.D. Acton et al., *Phys. Lett.* B **294** (1992) 436; OPAL Collaboration, *A Determination of $\sin^2\theta_W$ from an Inclusive Sample of Multihadronic Events*, OPAL Physics Note PN195 (1995)

196. LEP/SLD Electroweak and Heavy Flavour Working Groups, *A Combination of Preliminary Electroweak Measurements and Constraints on the Standard Model*, CERN-PPE/97-154 (1997)

197. *Physics at LEP 2*, CERN 96-01 (1996), eds. G. Altarelli, T. Sjöstrand, F. Zwirner

198. A. Ballestrero et al., *J. Phys.* G **24** (1998) 365

199. Takayuki Saeki, W Mass Measurement at LEP, talk presented at the XXXIVth Recontres de Moriond, Les Arcs, France, March 1999

200. ALEPH Collaboration, *Phys. Lett.* B **453** (1999) 121; ALEPH Collaboration, *Measurement of the W Mass in e^+e^- Collisions from 161 to 189 GeV*, ALEPH 99-017 CONF 99-012

201. DELPHI Collaboration, P. Abreu et al., *Eur. Phys. J.* C **2** (1998) 581; DELPHI Collaboration, *Measurement of the Mass of the W Boson Using Direct Reconstruction*, DELPHI 99-41 CONF 240

202. L3 Collaboration, M. Acciarri et al., *Phys. Lett.* B **454** (1999) 386; L3 Collaboration, *Preliminary Results on the Measurement of Mass and Width of W Boson at LEP*, L3 note 2377

203. OPAL Collaboration, G. Abbiendi et al., *Phys. Lett.* B **453** (1999) 138; OPAL Collaboration, *Measurement of the W boson Mass in e^+e^- Collisions at 189 GeV*, OPAL Physics Note PN385 (1999)

204. E. Gross, A.L. Read, D. Lellouch, *Prospects for the Higgs Boson Search in e^+e^- Collisions at LEP 200*, CERN-EP/98-094 (1998)

205. ALEPH Collaboration, *Phys. Lett.* B **447** (1998) 336; ALEPH Collaboration, *Search for the Neutral Higgs Bosons of the Standard Model and the MSSM in e^+e^- Collisions at 188.6 GeV*, ALEPH 99-007 CONF 99-003

206. DELPHI Collaboration, P. Abreu, *Search for Neutral Higgs Bosons in e^+e^- Collisions at $\sqrt{s} = 183$ GeV*, CERN-EP/99-06, accepted by *Eur. Phys. J.* C DELPHI Collaboration, *Search for Neutral Higgs Bosons in the Standard Model and the MSSM at 189 GeV*, DELPHI 99-8 CONF 208

207. L3 Collaboration, M. Acciarri et al., *Phys. Lett.* B **431** (1998) 437; L3 Collaboration, *Standard Model Higgs Searches at $\sqrt{s} = 189$ GeV*, L3 note 2382

208. OPAL Collaboration, G. Abbiendi et al., *Eur. Phys. J.* C **7** (1999) 407; OPAL Collaboration, *Search for Neutral Higgs Bosons in e^+e^- Collisions at $\sqrt{s} =$ 189 GeV*, CERN-EP/99-096, submitted to *Eur. Phys. J.* C

209. LEP Higgs Working Group, *Searches for Higgs Bosons: Preliminary Combined Results from the four LEP Experiments at $\sqrt{s} = 189$ GeV*, contributed paper to EPS-HEP-99, Tampere, 6-83

210. CDF Collaboration, F. Abe, *Phys. Rev.* D **50** (1994) 2966

211. D0 Collaboration, E. Barberis, CDF Collaboration, W. Yao, in: *Proceedings of the XXIX International Conference on High Energy Physics*, Vancouver 1998, eds. A. Astbury, D. Axen, J. Robinson, World Scientific, Singapore 1999

212. A. Heinson, CDF/D0 top physics, talk presented at the XXXIVth Recontres de Moriond, Les Arcs, France, March 1999

213. UA1 Collaboration, G. Arnison et al., *Phys. Lett.* B **122** (1983) 103

214. D0 Collaboration, B. Abbot et al., *Phys. Rev.* D **58** (1998) 092003

215. M. Lancaster, TEVATRON W mass, talk presented at the XXXIVth Recontres de Moriond, Les Arcs, France, March 1999

216. E.A. Paschos, L. Wolfenstein, *Phys. Rev.* D **7** (1973) 91

217. G.L. Kane, J.D. Wells, *Phys. Rev. Lett.* **76** (1996) 4458

218. Y. Zhu, Measurement of the Total Hadronic Cross Section in the $\sqrt{s} = 2\text{--}5$ GeV Range at BES, talk presented at the EPS-HEP-99, Tampere

Index

Springer Tracts in Modern Physics